Lecture Notes in Electrical Engineering

Volume 78

Alexander Biedermann and H. Gregor Molter
(Eds.)

Design Methodologies for Secure Embedded Systems

Festschrift in Honor of Prof. Dr.-Ing. Sorin A. Huss

Alexander Biedermann
Technische Universität Darmstadt
Department of Computer Science
Integrated Circuits and Systems Lab
Hochschulstr. 10
64289 Darmstadt, Germany
E-mail: biedermann@iss.tu-darmstadt.de

H. Gregor Molter
Technische Universität Darmstadt
Department of Computer Science
Integrated Circuits and Systems Lab
Hochschulstr. 10
64289 Darmstadt, Germany
E-mail: molter@iss.tu-darmstadt.de

ISBN 978-3-642-16766-9 e-ISBN 978-3-642-16767-6

DOI 10.1007/978-3-642-16767-6

Library of Congress Control Number: 2010937862

Typeset: Scientific Publishing Services Pvt. Ltd., Chennai, India.

Printed on acid-free paper

9 8 7 6 5 4 3 2 1

springer.com

Sorin A. Huss

Preface

This Festschrift is dedicated to Mr. Sorin A. Huss by his friends and his Ph.D. students to honor him duly on the occasion of his 60th birthday.

Mr. Sorin A. Huss was born in Bukarest, Romania on May 21, 1950. He attended a secondary school with emphasis on mathematic and scientific topics in Dachau near Munich and after his Abitur he studied Electrical Engineering with the discipline information technology at the Technische Universität München.

1976 he started his career at this university as the first research assistant at the newly established chair for design automation. Due to his very high ability he was a particularly important member of the staff especially in the development phase. In his research Mr. Huss dealt with methods for design automation of integrated circuits. The results of his research activities and his dissertation "Zur interaktiven Optimierung integrierter Schaltungen" were published in very high-ranking international proceedings and scientific journals. In his dissertation he started from the recognition that the computer-aided dimensioning of integrated circuits on transistor level normally leads to a very bad conditioned optimization problem and that this aspect played the central role in solving this problem. Mr. Huss provided important contributions to this matter which were advanced in future activities of the chair and finally resulted in the establishment of a company. Today, 15 employees of the company MunEDA are busy with the production and the world-wide sale of software tools for the design of analog components in microchips.

In 1982, Mr. Huss changed from university to industry and then worked at the AEG Concern in several positions. At last he was responsible for the development and adoption of new design methods as well as for the long-term application of the corresponding design systems as a department manager at the development center Integrated Circuits. Design tools that were developed under his direction, were used not only in the AEG Concern but also in famous domestic and foreign companies for the development of microelectronic circuits and systems. Important cross-departmental functions and the preparation of publicly funded major research projects indicate that his professional and organizational skills were well appreciated. Despite his technical-economical aim and the surrounding circumstances of internal release procedures Mr. Huss was able to document the academic level of the activities in his field of work by publications and talks outside of the company, too.

One of these publications brought forth the award of the ITG in 1988, for one of the best publications of the year. After six years of university experience and eight years of employment in industry, Mr. Huss had proved to be a well appreciated and internationally accepted expert in the field of the computer-aided design of integrated circuits and systems.

After having obtained a call for a C4 professorship in Computer Engineering at the Technische Hochschule Darmstadt, Mr. Huss started his work as a full professor in Darmstadt on July 1, 1990. Since that time Prof. Huss not only had a decisive impact on the technical sector of the Department of Computer Science. Since 1996 he acts as a co-professor at the Department of Electrical Engineering of the Technische Universität Darmstadt. With his assistance, the field of study Information System Technology (IST) was founded as cooperation between the Department of Computer Science and the Department of Electrical Engineering. In the same year, he rejected an appointment for a C4 professorship for technical computer science at the University Bonn and an offer as a head of the Institute for systems engineering at GMD, St. Augustin, to continue research and teaching at the TU Darmstadt.

On the basis of design methods for embedded systems, the focus of his research has enlarged and now connects aspects of heterogeneous systems with IT-systems and the automotive sector. More than 140 publications evidence his research activities. His contributions to research were acknowledged inter alia in 1988 with the Literature Award of the Information Technology Society (VDE/ITG), the Outstanding Paper Award of the SCS European Simulation Symposium in 1998, and both the Best Paper Award of the IEEE International Conference on Hardware/Software Codesign-Workshop on Application Specific Processors and the ITEA Achievement Award of the ITEA Society in 2004.

Apart from his memberships in ACM, IEEE, VDE/ITG and edacentrum he is – due to his expert knowledge about the design of secure embedded systems – head of one of three departments of the Center of Advanced Security Research Darmstadt (CASED). CASED was established in the year 2008 by the Hessian campaign for the development of scientific-economical excellence (LOEWE) as one of five LOEWE-Centers. Just the research group lead by Prof Huss which deals with the design of secure hardware, has presented more than fifty international publications since then. An Award for Outstanding Achievements in Teaching in the year 2005 and other in-house awards for the best lecture furthermore evidence his success in teaching. More than a dozen dissertations that have been supported by Prof. Sorin A. Huss to the present day complete the view of an expert, who does not only have extensive knowledge in his fields of research but is also able to convey his knowledge to others.

We wish Prof. Dr.-Ing. Sorin Alexander Huss many more successful years!

November 2010

Kurt Antreich
Alexander Biedermann
H. Gregor Molter

Joint Celebration of the 60th Birthdays of Alejandro P. Buchmann, Sorin A. Huss, and Christoph Walther on November 19th, 2010

Honourable Colleague,
Dear Sorin,

Let me first send my congratulations to you as well as to the other two guys. You jointly celebrate your sixtieths birthdays this year. I wish you all the best for your future work at TU Darmstadt! Sorin, let me first thank you for the time you served as one of my Deputies General in that treadmill they call "Dekanat". It was difficult for you to reserve some of your spare time for that voluntary, additional job. I appreciate that you agreed to take over this purely honorary post. Your advice has always been helpful for me. Your research and teaching activities are extraordinarily successful. Your success has led to a number of awards, from which I can only mention the most outstanding ones: Literature Award of the IT Chapter of the VDE (the German association of electrical engineers), European ITEA Achievement Award, Teaching Award of the Ernst-Ludwigs-Hochschulgesellschaft. You were one of the initiators of TU Darmstadt's bachelor/master program in information systems technology. To conclude this list, I would also like to mention that you are one of the domain directors of CASED, which is an important position not only for CASED itself but for the department and for TU Darmstadt as well. You three guys are true institutions of the departments (I am tempted to speak of dinosaurs, however, in an absolutely positive sense). You have seen colleagues come and go. Due to your experience and your long time of service in the department, you have become critical nodes of the departments corporate memory network. Your experience has been deciding many discussions typically (yet not exclusively) for the better. I should mention that each of you three guys is equipped with a specific kind of spirit. Your humorous comments, always to the point, made many meetings of the colleagues really enjoyable for the audience (well, the meeting chair did not always enjoy, but thats fine). You have always combining passion with reason, spirit with analysis, vision with rationality. On behalf of the colleagues, the department, and TU Darmstadt, I wish you three guys that you will have another great time together with all of us and an even longer chain of success stories than ever. Happy Birthday!

November 2010

Karsten Weihe
Dean of the Department of Computer Science
Technische Universität Darmstadt

The Darmstadt Microprocessor Practical Lab Some Memories of the E.I.S. Times

At the beginning of the 1980s, the publication of the book "Introduction to VLSI-Systems" by Mead/Conway initiated a revolution in the design of integrated circuits not only in the United States – it had a great feedback also in Germany. So it was intensively thought about establishing the design of integrated circuits as a field of study at the Technical Universities and Universities of Applied Sciences. Funded by the German Federal Ministry of Research and Technology (BMFT) the project E.I.S. (Entwurf Integrierter Schaltungen [design of integrated circuits]) was started in close cooperation with industrial concerns in 1983. The project was coordinated by the Society for Mathematics and Data Processing at Bonn (today: Fraunhofer Institute) and had the following objectives:

- Intensification of the research in the field of the design of microelectronic circuits at the Universities and progress in the theory of design methods
- Design and development of experimental CAD-software for microelectronic circuits for the use in research and teaching
- Design and test of application-specific circuits
- Enhancement of the number of computer scientists and electrical engineers with a special skill in VLSI-Design

Following the publication of the book "Introduction to VLSI-Systems" by Mead/Conway also people in Germany quickly recognized that VLSI-Design was not a kind of black magic but was based on a well-structured methodology. Only by means of this methodology it would be possible to handle the exponentially increasing design complexity of digital (and analog) circuits expected in the future.

Mr. Huss met this challenge very early and established a design lab at Darmstadt. The practical lab was intended to impart the complete design process from the behavior-oriented design model to the point of the layout. Based on a high-level-design methodology – which was taught in an accompanying lecture – and by using a design example, the entire development should be comprehended in detail and realized by means of modern CAE-tools. The practical lab therefore allowed an integrated education in the field of high-level-design methodology which was theoretically sound and deepened in practice. The design was carried out according to the principle of "Meet in the Middle" which was common practice instead of using the "Top Down" method. VHDL was used as formal language, as it enabled a description on all levels of abstraction. The necessary transformations were explained didactical cleverly using the Y-diagram of Gajski.

The main objective of the practical lab at this was not only learning the language, but to rehearse the Methodology in detail up of the design of a standard cell in a 1.5 m CMOS-technology with about 16,000 transistors. The production took place within the framework of the EUROCHIP-program. The abstract of the practical lab at the TH Darmstadt was introduced to an international group of experts at the 6th E.I.S. Workshop 1993 in T"ubingen. As the Goethe University Frankfurt ran a practical lab with a similar intention, a lively exchange of experiences arose subsequent to the E.I.S. Workshop. This fruitful cooperation in teaching later brought forth a textbook with the title "Praktikum des modernen VLSI-Entwurfs". The authors were Andreas Bleck, Michael Goedecke, Sorin A. Huss and Klaus Waldschmidt. The book was published by Teubner Verlag, Stuttgart in 1996. Unfortunately, it is no longer available due to developments in the publishing sector.

I gladly remember the cooperation with colleague Mr. Huss and his team. This cooperation later also continued in the area of research, for instance in the context of the SAMS-Project, which was funded by the BMBF and the edacentrum. For the future, I wish colleague Mr. Huss continued success and pleasure in teaching and research.

November 2010

Klaus Waldschmidt
Technische Informatik
Goethe Universität Frankfurt

Table of Contents

Towards Co-design of HW/SW/Analog Systems

Christoph Grimm, Markus Damm, and Jan Haase

Vienna University of Technology
Chair of Embedded Systems
Gußhausstrae 27-29
1040 Wien, Austria
{grimm,damm,haase}@ict.tuwien.ac.at

Abstract. We give an overview of methods for modeling and system level design of mixed HW/SW/Analog systems. For abstract, functional modeling we combine Kahn Process Networks and Timed Data Flow Graphs. In order to model concrete architectures, we combine KPN and TDF with transaction level modeling. We describe properties and issues raised by the combination of these models and show how these models can be used for executable specification and architecture exploration. For application in industrial practice we show how these models and methods can be implemented by combining the standardized SystemC AMS and TLM extensions.

Keywords: Analog/Digital Co-Design, KPN, Timed Data Flow, System Synthesis, Refinement, Refactoring, SystemC AMS extensions.

1 Introduction

Applications such as wireless sensor networks, cognitive radio, and multi-standard communication systems consist of multi-processor hardware, complex multi-threaded software, and analog/RF subsystems. A new complexity raised by such applications is the tight functional interaction between the different domains, even at mixed levels of abstraction. Therefore, specification and architecture level design require a comprehensive approach for system level design. System level design includes the following issues:

1. *Executable specification* of the intended behavior including analog/RF behavior and multi-process HW/SW systems.
2. *Architecture exploration* by mapping the executable specification to abstract processors, and adding SW that improves behavior of analog/RF components (calibration, error detection/correction, etc.).
3. *System integration*, mostly by mixed-level simulation, upon availability of hardware designs and software programs.

Compared with HW/SW co-design, the co-design of HW/SW/Analog systems lacks models, methods and tools that go beyond modeling and simulation. A major problem for co-design of HW/SW/Analog systems is that modeling and

A. Biedermann and H. Gregor Molter (Eds.): Secure Embedded Systems, LNEE 78, pp. 1–24.
springerlink.com

design of HW/SW systems at one hand, and of analog systems at the other use fundamentally different methods:

- HW/SW Co-design is done usually "top-down", relying on existing platforms that enable to some extent abstraction from realization. In contrast, analog design is rather done "bottom up".
- Design of HW/SW systems can to some extent be automated and formalized. In contrast, analog design is sometimes considered as "black magic".

In this work we give an overview of methods that together draw the vision of a co-design methodology that is applicable to HW/SW/Analog Systems as a whole as shown by Fig. 1. We simplify the problem a lot by taking the analog (and also digital) circuit design out of the challenge. Instead we assume that "HW/SW/Analog Co-Design" gets characterized models from analog design, and validated IP or platforms from digital design. Like in HW/SW Co-Design, we propose an interactive strategy where architecture mapping selects a limited number or architectures that are evaluated by modeling and simulation based on SystemC. However, we propose to restrict modeling techniques to Kahn Process Networks (KPN), Timed Data Flow (TDF), and Transaction Level Modeling (TLM). This allows us to also address issues such as (maybe in future work automated) partitioning or system synthesis.

In the following, we first describe related work and KPN, Timed Data Flow (TDF) and Transaction Level Modeling (TLM). In Sect. 2 we discuss issues

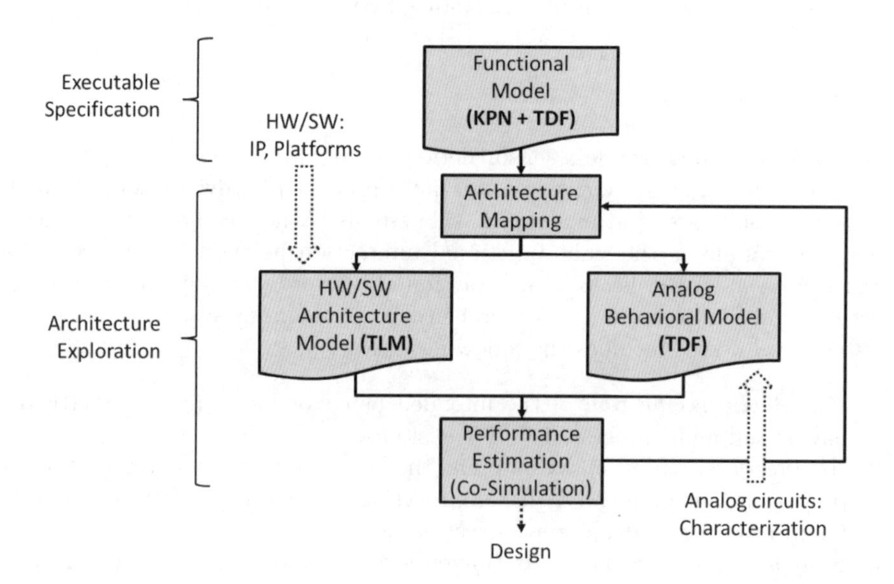

Fig. 1. HW/SW/Analog Co-Design with Executable Specification and Architecture Exploration consisting of architecture mapping and performance estimation by co-simulation of mixed TLM/TDF models

raised by the combination of KPN, TDF and TLM and show how they can be modelled using SystemC. In Sect. 3 we show how the combination of KPN, TDF and TLM models can be used for architecture exploration. In Sect. 4 we discuss a real-world example.

1.1 Related Work

For many years, modeling languages were the main tools for designers of HW/-SW/analog systems. A prominent success was the advent of standardized and agreed modeling languages such as VHDL-AMS [1] and Verilog-AMS [2] that focus design of analog/digital subsystems. Recently, AMS extensions for SystemC have been standardized that address the co-design of mixed HW/SW/Analog systems better [3]. However, we believe that co-design should be more ambitious than just being able to co-simulate. First attempts to address co-design are made in [4,5,6,7] focusing the design of analog/mixed-signal subsystems. Interactive methodologies that tackle the transition from data flow oriented descriptions to analog/digital circuits are described in [8,9,10,11]. In order to deal with overall HW/SW/Analog systems, models from HW/SW Co-Design and from analog/RF design have to be combined.

In *HW/SW Co-Design*, models must show a maximum of parallelism. Task Graphs [12], Kahn Process Networks (KPN) [13] or Synchronous Data Flow (SDF) [14] are widely used in that context. Especially KPN and SDF maintain a maximum of parallelism in the specification while being independent from timing and synchronization issues, thus being useful for executable specification. Pioneering work was done in the Ptolemy Project [15,16]. Jantsch [17] gives a good summary and formalizes models of computation applied in embedded system design, including combined models. Transaction Level Modeling (TLM, [18,19]) and other means in SpecC and SystemC in contrast specifically enable to model timing and synchronization at architecture level. SysteMoC enables design of digital signal processing systems combining several models [20], but lacks support for analog/RF systems.

In *analog/RF systems*, behavioral representations abstract from physical quantities. However, abstraction from time is hardly possible because typical analog functions such as integration over time are inherently time dependent. Block diagrams in Simulink or Timed Data Flow (TDF) in the SystemC AMS extensions [3] therefore abstract structure and physical quantities while maintaining continuous or discrete time semantics.

It is difficult to bring the worlds of HW/SW Co-Design and analog/RF design together. Main focus of the Ptolemy project [15] was simulation and HW or SW synthesis, but not overall system synthesis. Approaches such as Hybrid Automata ([32], Model Checking) lack the ability to deal with complex HW/SW systems. Hybrid Data Flow Graphs (HDFG [5,21], Partitioning) focus the border between discrete and continuous modeling. The functional semantics of HDFG offers – in combination with functional languages for system specification – interesting perspectives for specification of parallel systems while being able to describe HW/SW/analog systems. However, due to availability of tools and languages we focus on KPN and TDF in the following as a starting point for system level synthesis, and TLM for architecture level modeling.

1.2 Kahn Process Networks, Timed Data Flow, and TLM

Kahn Process Networks (KPN). KPN are a frequently used model of computation that allows easy specification of parallel, distributed HW/SW systems. In KPN, processes specified e.g. in C/C++ communicate via buffers of infinite length. Writing is therefore always non-blocking, whereas reading is blocking. KPN are an untimed model of computation: Timing is not specified and not necessary because the results are independent from timing and scheduling. KPN are specifically useful for the executable specification of HW/SW systems, because they – in contrast to sequential program languages – maintain parallelism in an executable specification and therefore enable the mapping to parallel hardware, e.g. multi-processor systems. A particular useful property of KPN is that it enables abstraction of timing and scheduling: Outputs only depend on the input values and their order (*determinacy*), provided all processes are deterministic. Non-determinacy can for example be introduced by a non-deterministic merge process.

In order to enable execution, scheduling algorithms (e.g. Park's algorithm [22]) may be defined that restrict the size of the buffers (Bounded KPN, BKPN). However, a limited size of buffers cannot be guaranteed in general.

Timed Data Flow (TDF). In order to overcome the restrictions of KPN considering scheduling and size of buffers, different subsets of KPN have been defined, most prominent of them the Synchronous Data Flow (SDF, [14]). In SDF, an undividable execution of a process consumes a constant number of tokens or samples from the inputs and generates a constant number of tokens at the outputs. Under these conditions, a static schedule with size-limited buffers can be determined before execution of the processes by solving the balancing equations for a cluster of processes. For repeated inputs, the schedule is repeated periodically. Like KPN, SDF is an untimed model of computation. Nevertheless, SDF is used for representing digital signal processing (DSP) methods, assuming constant time steps between samples.

In Timed Data Flow (TDF, [3,31]), each process execution is assigned a time step. In case of multiple samples per execution, the time step is distributed equally between the samples. Apart from specification of DSP algorithms, this enables the representation of analog signals by a sequence of discrete-time samples while abstracting from physical quantities and assuming a directed communication between analog components. A major benefit of TDF for the specification of analog and RF systems is the ability to embed other "analog" formalisms in the processes. The analog formalisms can be transfer functions

$H(z)$ or $H(s)$ and models of computation such as signal flow graphs or electrical networks while maintaining advantages of SDF such as finite sizes of buffers, static scheduling and analysis at compile-time. A limitation is the discrete-time representation of inputs and outputs.

Transaction Level Modeling (TLM) is rather a means for abstract modeling than a model of computation. It provides means to model communication of HW/SW systems at architecture level. The idea of Transaction Level Modeling (TLM) is to bundle the various signal-level communication events (request, acknowledge, ...) into a single element called a *transaction*. The OSCI TLM 2.0 standard [19] defines a TLM extension for SystemC. The TLM extensions for SystemC use a dedicated data structure for representing a transaction (the *generic payload*), which contains elements like a target address, a command and a data array. Transactions – or more precisely references to them – are then passed around through the system via method interfaces, and each system part works on the transaction depending on the role of the system part. A simple example might look like this: A TLM *Initiator* (e.g. a processor module) creates a transaction, sets its command to READ, sets its target address to the address in its virtual address space it wants to read from, and reserves an appropriate data array for the data to read. It then sends the transaction through a socket to a TLM *Interconnect* (e.g. a bus or a router), which might do arbitration, and passes it on to the appropriate target device according to the address, after mapping the virtual address to the physical address within the target device. The TLM *Target* (e.g. a memory) then would copy the desired data to the transactions' data array, and return the transaction reference.

This basically implements a memory mapped bus. However, in TLM 2.0 it is possible to augment the generic payload with custom commands. An important observation is that passing the reference through the system might look like passing a token, but the data associated to that token is mostly non-atomic and can in theory be of arbitrary size.

Regarding time, the TLM 2.0 standard offers interesting possibilities. By using *temporal decoupling*, it is possible to let every TLM initiator run according to its own local time, which is allowed to run ahead of the global simulation time. This is often referred to as *time warp*. The TLM 2.0 method interface foresees a delay parameter to account for the local time offset of the initiator issuing the transaction.

Synchronization with global time only occurs when needed, or with respect to predefined time slots. This technique reduces context switches and therefore increases simulation performance, with a possible loss of simulation accuracy. Using the facilities provided by the TLM 2.0 extensions in such a manner is referred to as the *loosely timed coding style*, in contrast to the so-called *approximately timed coding style*, where each component runs in lock-step with the global simulation time, and where the lifetime of a transaction is usually subdivided into several timing points.

2 Executable Specification

The executable specification of a HW/SW/Analog system must be able to capture both analog behavior and behavior of a HW/SW system. Analog behavior is usually modeled by transfer functions or static nonlinear functions, i.e. using TDF model of computation; HW/SW systems are specified by multiple threads or processes, i.e. KPN. In the following we first discuss the combination of KPN and TDF for specification of behavior of HW/SW/Analog systems. Then we show how to model behavior in these models of computation using SystemC.

2.1 Combination of KPN and TDF

A combination of KPN and TDF with embedded "analog" models of computation offers properties required for both HW/SW Co-Design and design of analog systems, thus enabling HW/SW/Analog Co-Design. However, we have to define semantics of the combined KPN/TDF model of computation carefully considering the following issues:

- KPN is an untimed model of computation, but TDF requires a concept of time. One way to overcome this heterogeneity is to introduce time in KPN. However, KPN with time extensions will no longer be determinate in the sense of outputs which are independent from timing and scheduling.
- TDF is activated by advancement of time, assuming that a defined number of input samples is available (guaranteed e.g. by schedule of KPN). However, with arbitrary scheduling and timing, KPN output could produce too many or not enough samples for TDF input.

In the following we examine communication from TDF to KPN and from KPN to TDF. The communication from TDF to KPN is unproblematic: As KPN are untimed, they can execute when there are samples at the inputs, regardless of the time when they are produced by TDF. However, the communication from KPN to TDF is a more complex issue: TDF processes are executed following a static schedule that is triggered by time. The static schedule guarantees that enough samples are available at connections between TDF processes. However, this is not valid for communication from KPN (with by definition arbitrary scheduling) and TDF processes. Two different extensions solve this issue in different ways:

1. Assuming a KPN schedule that in time generates enough samples at the inputs of TDF.
2. Merging "default token" in case of absence of inputs from KPN to the TDF processes, e.g. because 1 has failed.

In the first case, the property of determinacy is preserved for the whole model. In the second case, determinacy is not preserved because the default tokes are merged in a non-deterministic way.

Simple Example: Executable Specification of a Signal Generator. As an example we give a very simple signal generator shown by Fig. 2. The signal generator produces a saw tooth shaped signal. We use a TDF model to describe integration over time, and a KPN to observe the TDF output and to switch the input of the TDF depending on the limits *ul* and *ll*. The comparison is done by two KPN processes that continuously get a stream of TDF inputs, and omit an output *ub* or *lb* only if the upper or lower border is crossed.

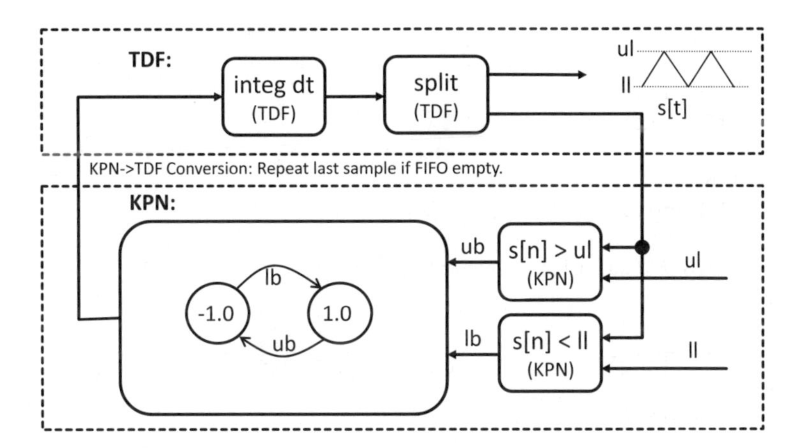

Fig. 2. A signal generator that generates saw tooth signals within limits ul and ll can be modeled by a KPN with three processes (one specified by an automaton) and a TDF of two nodes

The output of the KPN is generated by a process that waits on an input from either comparison, then switches its internal state, and produces an output in the same rate as needed by the TDF cluster. In the given example we assume the scheduling will ensure that the TDF gets always enough inputs and we do not need to insert a special converter that would insert default outputs in case the KPN processes are not scheduled in time.

2.2 Modeling and Simulation of KPN and TDF with SystemC

SystemC is a modeling language for HW/SW Co-Design that is based on C++. SystemC allows using C++ for the specification of processes. Processes are embedded in modules that communicate with each other via ports, interfaces, and channels. There are several interfaces and channels already defined in SystemC, but they can also be defined by the user. This way, various models of computation can be supported, including KPN (via the already existing `sc_fifo` channel) and SDF (by extending SystemC). Time in SystemC is maintained globally by a discrete event simulation kernel. In the recently introduced AMS extensions, the TDF model of computation is provided. In the following we show how SystemC can be used to specify a mixed HW/SW/Analog system within the semantics of combined KPN and TDF.

Modeling KPN using SystemC. Kahn Process Networks can be modeled by using processes (`SC_METHOD`) which communicate only via FIFOs (`sc_fifo<T>`) with appropriate properties: Blocking read, infinite size, and hence non-blocking write. Obviously, no implementation can offer infinite size, but in practice it is possible to have "large enough" FIFOs. In the case of a full FIFO, one can use blocking behavior and omit an error message. The following example shows a KPN process that distributes a stream of input samples u to two streams v and w:

```
SC_MODULE(split)
{
  sc_fifo_in<int>  u;
  sc_fifo_out<int> v, w;

  bool n;
  SC_CTOR(split): n(true)
  {
    SC_THREAD(do_split); sensitive << u;
  }

  void do_split()
  {
    while(true)
      {
       if(n) v.write( u.read() ); else w.write( u.read() );
       n = !n;
      }
  }
}
```

Modeling TDF using SystemC AMS extensions. The TDF model of computation is introduced by the AMS extensions 1.0 [3]. TDF models consist of TDF modules that are connected via TDF signals using TDF ports. Connected TDF modules form a contiguous structure called TDF cluster. Clusters must not have cycles without delays, and each TDF signal must have one source. A cluster is activated in discrete time steps. The behavior of a TDF module is specified by overloading the predefined methods `set_attributes()`, `initialize()`, and `processing()`:

- The method `set_attributes()` is used to specify attributes such as rates, delays or time steps of TDF ports and modules.
- The method `initialize()` is used to specify initial conditions. It is executed shortly before the simulation starts.
- The method `processing()` describes the time-domain behavior of the module. It is executed at each activation of the TDF module.

At least one definition of the time step value and, in the case of cycles, one definition of a delay value per cycle has to be done. TDF ports are single-rate by default. It is the task of the elaboration phase to compute and propagate consistent values for the time steps to all TDF ports and modules. Before simulation, the scheduler determines a static schedule that defines the order of activation of the TDF modules, taking into account the rates, delays, and time steps. During simulation, the `processing()` methods are executed at discrete time steps. The following example shows the TDF model of a mixer. The `processing()` method will be executed with a time step of 1 µs:

```
SCA_TDF_MODULE(mixer) // TDF primitive module definition
{
  sca_tdf::sca_in<double>  rf_in, lo_in; // TDF in ports
  sca_tdf::sca_out<double> if_out;       // TDF out ports
  void set_attributes()
  {
    set_timestep(1.0, SC_US);  // time between activations
  }
  void processing()                     // executed at activation
  {
    if_out.write( rf_in.read() * lo_in.read() );
  }
  SCA_CTOR(mixer) {}
};
```

In addition to the pure algorithmic or procedural description of the `processing()` method, different kind of transfer functions can be embedded in TDF modules. The next example gives the TDF model of a gain controlled low pass filter by instantiating a class that computes a continuous-time Laplace transfer function (LTF). The coefficients are stored in a vector of the class `sca_util::sca_vector` and are set in the `initialize()` method. The transfer function is computed in the `processing()` method by the ltf object at discrete time points using fixed-size time steps:

```
SCA_TDF_MODULE(lp_filter_tdf)
{
  sca_tdf::sca_in<double>  in;
  sca_tdf::sca_out<double> out;
  sca_tdf::sc_in<double>   gain;// converter port for SystemC input
  sca_tdf::sca_ltf_nd ltf;      // computes transfer function
  sca_util::sca_vector<double> num, den; // coefficients

  void initialize()
  {
    num(0) = 1.0;
    den(0) = 1.0;
    den(1) = 1.0/(2.0*M_PI*1.0e4); // M_PI=3.1415...
  }
  void processing()
  {
    out.write( ltf(num, den, in.read() ) * gain.read() );
  }
  SCA_CTOR(lp_filter_tdf) {}
};
```

The TDF modules given above can be instantiated and connected to form a hierarchical structure together with other SystemC modules. The TDF modules have to be connected by TDF signals (`sca_tdf::sca_signal<type>`). Predefined converter ports (`sca_tdf::sc_out` or `sca_tdf::sc_in`) can establish a connection to a SystemC DE channel, e.g. `sc_signal<T>`, reading or writing values during the first delta cycle of the current SystemC time step.

As discussed above, conversion from TDF to KPN is quite easy, if we assume an unbounded FIFO. A converter just has to convert the signal type:

```
{
  sca_tdf::sca_in<double>  in_tdf;       // TDF in-port
  sc_fifo_out <double> out_kpn, // KPN out-port
  void processing()
  {
    out_de.write( in_tdf.read() );
  }
  SCA_CTOR(tdf_kpn_converter) { }
};
```

However, as explained before, the conversion can lead to a situation where not enough samples from KPN are available. The following piece of code demonstrates a simple converter from KPN to TDF that inserts a "DEFAULT" value and emits a warning to the designer that determinacy is no longer a valid property of the combined KPN/TDF model.

```
SCA_TDF_MODULE(kpn_tdf_converter)
{
  sc_fifo_in<double> in_kpn,          // KPN in-port
  sca_tdf::sca_out<double> out_tdf;   // TDF out-port
  void processing()
  {
      if ( !in_kpn.empty() )             // FIFO empty?
      out_tdf.write( in_kpn.read() );
    else                              // we insert default
    {
      out_tdf.write( DEFAULT );       // DEFAULT from somewhere
      warning("Non-determinate due to merge with default values");
    }
  }
  SCA_CTOR(kpn_tdf_converter) { }
};
```

3 Architecture Exploration

For HW/SW/Analog Co-Design we need a transition from an executable specification to one or more architectures, for which quantitative performance properties such as accuracy or power consumption are estimated. We call this process

"Architecture Exploration". In the following, we describe first how to map the executable specification to concrete architectures. It has to be emphasized that the implementation modeled by such an architecture mapping should show similar behavior to the exploration – with the difference that communication/synchronization and timing/scheduling are modeled in more detail for allowing performance estimation by modeling/simulation. Then, we focus on modeling issues that are raised when combining SystemC AMS extensions for modeling analog behavior, and SystemC TLM 2.0 for modeling architecture of HW/SW systems.

3.1 Architecture Mapping to Analog and HW/SW Processors

In order to come to an implementation of the executable specification, we map processes of KPN and TDF to allocated processors, and channels between KPN or TDF processes to allocated communication/synchronization elements as depicted in Fig. 3.

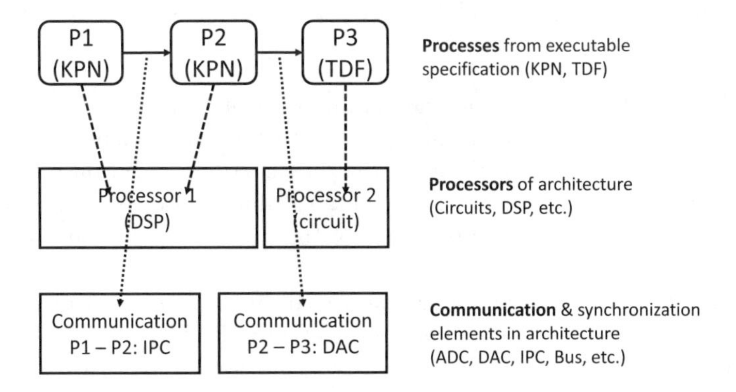

Fig. 3. Architecture mapping assigns processes to allocated processors and channels to allocated communication/synchronization elements that implement the specified behavior

Processors are components that implement the behavior of one or more processes. Communication/synchronization elements are components that implement the behavior of one or more channels.

TDF processes can be implemented by the following kind of processors:

- Analog subsystem. We consider an analog subsystem as an "analog processor" in the following. Analog processors cannot be shared; hence the relation between processors and processes is a 1:1 mapping in this case. The behavior of analog subsystems at architecture level can be modeled by adding the non-ideal behavior of the analog subsystem to the TDF model from the executable specification.
- DSP software running on a digital processor (DSP, Microcontroller, etc.). A DSP can implement a number of TDF processes, if the processor has appropriate performance, and a facility that takes care of discrete time scheduling

(e.g. interrupt service routing, ISR). For performance analysis, an instruction set simulator can replace the TDF processes implemented by the processor.

TDF clusters with feedback loops are a special case. The implementation depends on the delay that is inserted into the loop. If it is very small compared with time constants, the delay can be negotiated for implementation. If it is a dedicated delay used for specifying a delay in DSP methods, a switched capacitor or digital implementation is required.

KPN processes can be implemented by the following kind of processors:

- Software processes running on a digital processor. Such processors can, given an appropriate performance, implement one or more processes from one or more KPN. For performance estimation, an instruction set simulator can replace the TDF processes that are implemented by the processor.
- Application-specific co-processors. Such processors can, given an appropriate performance, implement one or more processes from one or more KPN. For performance analysis, the KPN processes can be substituted with timed models of the co-processors.

Special care has to be taken on the communication/synchronization elements. FIFOs of infinite size are obviously not an efficient implementation (although convenient for specification). For that reason the communication infrastructure in KPN and TDF is replaced with concrete communication/synchronization elements that implement similar behavior of the overall system with low cost for implementation.

The elements to be used for implementation depend on the mapping of processes to processors: Communication between KPN processes mapped on a micro-processor/DSP might use infrastructure for inter-process communication available in a real-time OS. However, in many cases an application-specific co-processor is used. In that case, one has to model communication between processor and (co-)processor at architecture level. Such communication usually goes at physical layer via memory-mapped registers or memory areas to which the communicating processes have access. Since this is a crucial part of the overall system (both considering correctness and performance), it was a focus of research in HW/SW Co-Design. Transaction-level modeling has become one of the most important means for modeling and simulation of this part. After architecture mapping, we get a model that consists of:

- TDF models that model the parts mapped to analog subsystems.
- An Instruction set simulator (ISS) that models the embedded SW parts and the used embedded platform hardware (DSP, Microcontroller). The ISS is interfacing either directly with the discrete-event simulator via TLM modeling.
- A TLM model that models the communication with specific co-processors that are modeled using the discrete-event model of computation

Obviously, many different mappings of an executable specification in KPN/TDF to processors are possible. Usually, only very few mappings are useful and often roughly known. In a number of cases the system partitioning in HW/SW/Analog systems can be determined by the following criteria:

1. Comparison operations are a good cut between analog and HW/SW domain if the compared values are in analog, because this saves a costly and slow A/D converter, e.g. like in the saw tooth generator example.
2. Signal conditioning of very small physical quantities, or at very high frequencies can only be implemented in analog, e.g. RF frontend in radio, instrumentation amplifier.
3. Clear hints for a digital implementation are: Signal processing in lower frequencies, high requirements for accuracy or for re-configurability, and processes with complex control-flow (e.g. automata).

In order to optimize the mapping optimization techniques that go across the borders of different models of computation are useful as described in [5,21] using HDFG. Fig. 4 shows the principle of such optimizations.

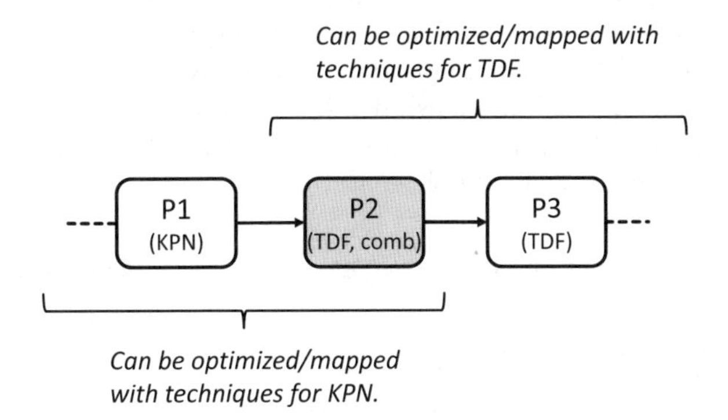

Fig. 4. Optimization techniques for KPN and TDF are applicable across the borders between these models of computation under the conditions described below (details: see [5])

For such optimizations and to enable a higher flexibility in architecture mapping – note, that KPN/TDF processes can also be split into several processes – processes have to be identified whose semantics are identical, no matter if they are executed following the TDF or the KPN rules. The following rules apply:

1. Processes with time dependent function such as integration over time, or transfer functions in TDF cannot be shifted to KPN without changing overall behavior, because the non-deterministic scheduling of KPN would have impact on time of execution and hence the outputs.

2. Processes with internal state cannot be shifted across converters between KPN and TDF that non-deterministically insert "default" samples without changing overall behavior, because the inserted tokens would have an impact on the state and hence, the outputs.
3. Processes with internal state and time-independent function state can be shifted across converters between KPN and TDF if scheduling ensures samples from KPN to TDF are available, because in this case KPN and TDF both are pure stream-processing functions.
4. Processes with combinational function can be shifted between from KPN to TDF and vice versa, as long as the number and order of inserted default samples is not changed.

Simple Example: Architecture Mapping Signal Generator Specification. As an example we use again the simple signal generator. Of course, the KPN and TDF nodes can be mapped to (HW) processors that are close to their model of computation. The integrator in the TDF section, for example, can be implemented by an analog circuit in an easy and efficient way, while the KPN processes could be implemented using a microcontroller. The communication between microcontroller (with KPN implementation) and the analog integrator (with TDF implementation) will then require an analog/digital converter (ADC) that produces a discrete-time stream of outputs. We get another, more efficient mapping by "shifting" the border between KPN over the two comparisons $s[n] > ul$ and $s[n] < ll$ as shown in Fig.5. In this case we can assume that these two (initially KPN) processes become part of the TDF cluster, and hence we can also allocate an analog implementation for these processes. Because the

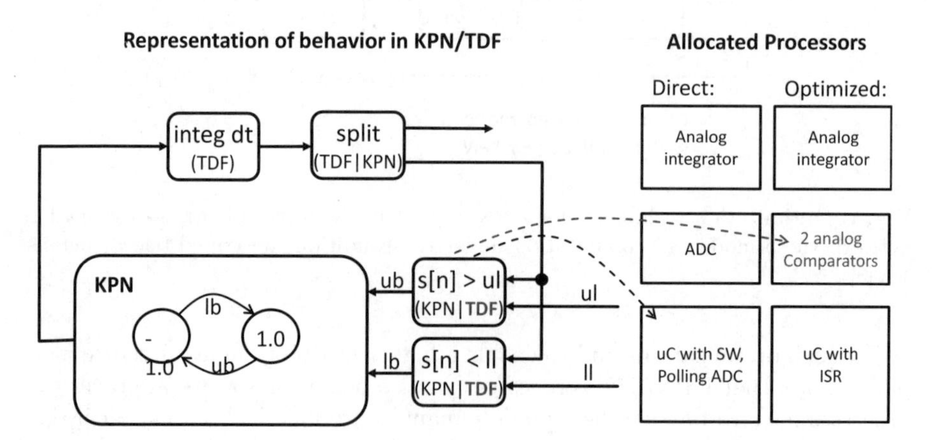

Fig. 5. Application of Repartitioning on the simple example allows finding a more appropriate architecture mapping. The nodes that observe the output ($s[n]$) are better implemented in continuous/discrete-time hardware that by a software that would have to poll the output of an analog/digital converter (ADC).

output of a comparison is digital, no ADC is required. The two comparators can be used to trigger interrupts that start an interrupt service routing that executes the automaton from the KPN.

Unfortunately, the most efficient architecture isn't as obviously to find as in the simple example given. For performance estimation in complex HW/SW/Analog Co-Design we have to create models that combine TDF models describing the analog part, an ISS (e.g. with TLM interfaces), and TLM models that models the HW/SW architecture and the operating system.

3.2 Estimation of Quantitative Properties by System Simulation

For comparison of different architectures quantitative properties of an implementation have to be estimated. Known from HW/SW systems are estimation methods for the properties chip area, throughput, (worst case) execution time (e.g. [26]), and power consumption (e.g. [23,28]). Power profiling for complex HW/SW/Analog systems such as WSN is described in [25,27]. Although these properties are also to be considered for partitioning of HW/SW/analog systems, additional parameters are needed especially due to analog/RF subsystems. Accuracy and re-configurability are the most important criteria for partitioning if analog implementation is considered as well.

Accuracy is an important property because analog circuits have initially low accuracy that is insufficient for most applications. The art of analog design is to make an inaccurate circuit accurate by error cancellation techniques such as calibration, feedback or correlation, often involving digital signal processing. Estimation of accuracy therefore requires means to estimate accuracy across the border between discrete and continuous signal processing. First attempts in this direction have been done using a kind of automated error propagation computation by affine arithmetic [33].

Re-Configurability is a property that is to some extend only offered by HW/-SW implementation, but not only limited by an analog/RF implementation. Hence, the need for full re-configurability of a process requires its implementation in software. Reduced requirements for re-configurability of functionality only on occasion allow also FPGA coprocessor architectures. If re-configurability is restricted to changing parameters, or selecting a small number of alternatives, an analog implementation is possible. Partitioning strategies that specifically take care of re-configurability are described in [21].

Although the above mentioned estimation methods give some hints for improved partitioning, an overall system simulation at architecture level is absolutely essential due to complexity of most properties. Architecture level simulation should include models of all non-ideal properties of subsystems that have impact on the overall system behavior. To get architecture level models efficiently from a functional model such as the executable specification, methods from SW engineering such as refinement [11] or refactoring [9,10] are useful.

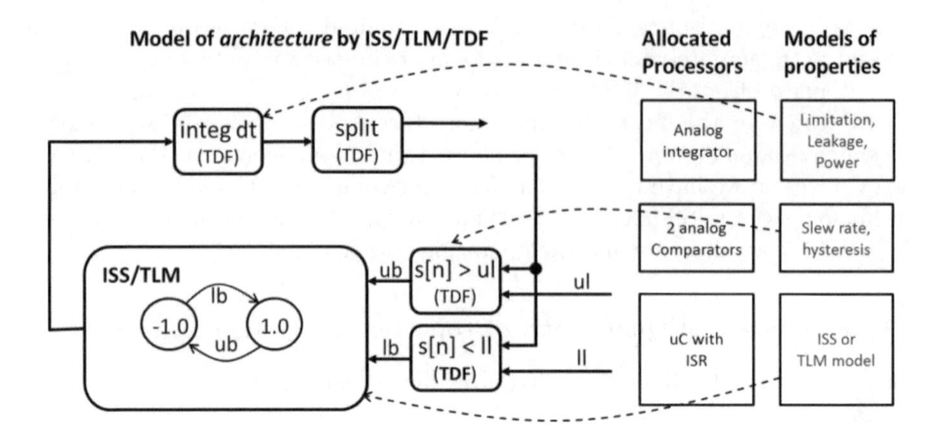

Fig. 6. Refinement of the processes from the executable specification to a model of an architecture by using models of non-ideal properties e. g. from the TU Vienna building block library

Refinement as proposed in [11] is based on a library with models of non-ideal effects (e.g. [29]) such as jitter, or noise that can be used to augment the ideal, functional model with non-ideal properties for evaluation of different architectures while maintaining the abstract modeling using TDF. For example, we can model sampling jitter in the TDF model of computation by assuming an initially ideal signal $x(t)$ that is sampled with a small delay δt caused by jitter. Hence, we assume that

$$\frac{x(t+\delta t) - x(t)}{\delta t} \approx \frac{dx}{dt} \tag{1}$$

and the error from jitter $\in_{jitter}$ as $x(t+\delta t) - x(t) \approx \delta t dx/dt$. Hence, we can refine the ideal signal $x(t)$ to a signal that is sampled with a jittering signal by adding $\in_{jitter}$ to the ideal signal. In a similar way effects for most typical architecture level effects (distortions of nonlinear switches, limitation, noise, etc.) are modeled in the TUV library for communication system design.

Use of refactoring is proposed in [9,10]. Refactoring techniques are well suited to support the transition from abstract modeling in block-diagrams (e.g. in TLM model of computation) towards modeling with physical quantities and electrical networks by automatically changing the interfaces of the models.

3.3 Coupling of TDF and TLM Models of Computation

For estimation of quantitative performance parameters such as accuracy (bit or packet error rates), a high-performance overall system simulation at architecture level is required. Behavioral models of analog circuits can be modeled easily using the TDF model of computation. For modeling performance of HW/SW architectures, loosely and approximately timed TLM models are useful. Therefore, a co-simulation between TDF and TLM model of computation is required. An important aspect of TDF/TLM co-simulation is how to integrate TDF data streams into TLM simulation and how to provide data streams out of TLM to TDF. In the following we describe how co-simulation between a TDF model and a *loosely* timed TLM model can be achieved. In loosely timed TDF models, the TLM initiators involved might run ahead of simulation time. Regarding the data, the approach is straightforward:

- **TLM to TDF:** The data bundled into the (write-) transactions arrives irregularly; therefore it has to be buffered into a FIFO. The TDF process then reads from this FIFO.
- **TDF to TLM:** The streaming data from the TDF side arrives regularly; yet the (read-) transaction requesting this data are irregular in general regarding time and data size requested. Here, a FIFO is also the solution: the TDF process writes on it, and it is emptied by the TLM side.

Of course, like in the coupling of KPN and TDF, corner cases can occur if the FIFOs run empty or (since we assume bounded FIFOs here) full. It is up to the designer how these cases are handled, since the semantics depend on the model at hand. If it is not acceptable, errors can be thrown, but providing fallback values resp. discarding values might also be an option.

To minimize these corner cases, proper synchronization is very important. Since we assume loosely timed initiators on the TLM side, we can exploit that TDF (in the way it is implemented in the SystemC AMS extensions) also in general runs ahead of the global simulation time (especially if large data rates are used). That is, $t_{TDF} \geq t_{DE}$ always holds, where t_{TDF} denotes the SystemC AMS simulation time, and t_{DE} denotes the SystemC simulation time.

Figure 7 illustrates this with a TDF module, which consumes 3 data tokens every 2 ms, and produces 2 data tokens every 3 ms. Beneath the tokens the point in (absolute) simulation time, where the respective token is valid, is indicated. Above and below that, the values for t_{TDF} and t_{DE} are given, when the tokens are processed.

At synchronization points, e.g. by accessing SystemC AMS converter ports that synchronize the SystemC discrete event (DE) kernel with TDF, the SystemC AMS simulation kernel yields to the SystemC simulation kernel such that the DE processes can catch on.

Figure 7 also shows that further time warp effects result from using multi-rate data flow. When a TDF module has an input port with data rate >1 it also receives "future values" with respect to tDE, and even tTDF . When a TDF module has an output port with data rate >1, it also sends values "to the future" with respect to t_{DE} and t_{TDF} . The difference to TLM is that the effective local time warps are a consequence of the static schedule, with the respective local time offsets only varying because of interrupts of the static schedule execution due to synchronization needs.

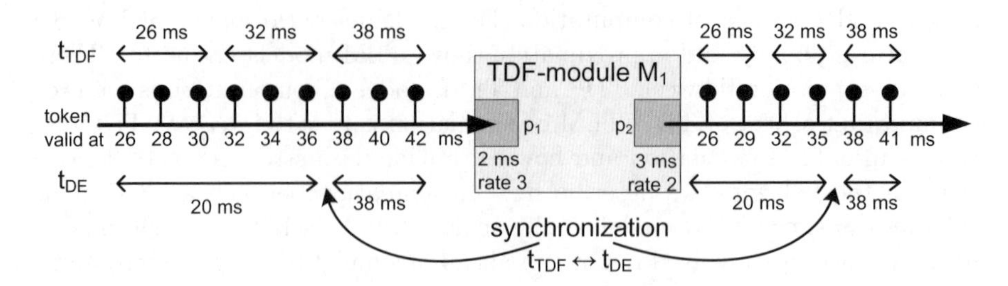

Fig. 7. Communication between TDF and TLM models of computation has to consider the time defined by the DE simulation kernel and the time that can run ahead in TDF and in TLM

This shows that with respect to the loosely timed TLM initiators, we have to deal with two parties which run ahead of global simulation time in their own special way. The question is when to trigger synchronization. In general, we want to trigger synchronization as infrequent as possible. But what are the conditions for triggering synchronization? The answer lies in the corner cases described above (buffer over- resp. underflow). We use these corner cases to trigger synchronization:

- **TLM to TDF:** If a buffer runs full, the transaction can simply be returned with an error message (a mechanism foreseen in TLM 2.0). If a buffer runs empty, however, we still have to feed the unstoppable TDF side. By synchronizing now we give the TLM side the chance to "catch on" and provide sufficient write transaction to fill the buffer.
- **TDF to TLM:** Again, an empty buffer in this case can be handled by a TLM error. A full buffer, however, needs new read-transactions to free space, which might be provided due to synchronization.

Of course, synchronization might still not resolve the respective issue. Then the designer has to specify the desired corner case behavior as described above.

The TDF to TLM conversion direction could be handled in an even more straightforward manner, if we would allow the converter (i.e. the TDF side) to act as an TLM initiator. In this case, the delay of the transaction could be set to the difference of tDE and the valid time stamp of the last token sent via the transaction (i.e. t_{TDF} + token number $\cdot$ sampling period). Note that in this active approach there is no need for synchronization due to this delay annotation.

However, such an initiator would be very limited, since there is no natural way to obtain other transaction data like target address or command out of the streaming TDF input. Yet, it can make sense for a TLM initiator using TDF processes, for example when modeling a DSP which uses TDF models of signal processing algorithms. In this case, we have to write an application specific converter.

A possible approach for such a converter would work by dividing the TDF to TLM conversion into two parts incorporating the converter described above and using an additional wrapper, which acts on top of the "raw" converter and implements the necessary semantics (see also Sect. 2.1). In general, the wrapper would do two things: mutating transactions by interpreting their data sections, or creating new transactions.

A DSP model might get the request for a certain computation, which might be encoded by a transaction using extensions, as it is foreseen by the TLM 2.0 standard. It then would create a raw data transaction and send it (via a converter) to the appropriate TLM model performing the necessary algorithm, and insert the response data to the original transaction and return it. Another scenario might involve a raw data transaction carrying the target address encoded within its data section.

Then the wrapper would generate the new transaction target address accordingly, and delete the encoded address out of the data section. Note that the basic converters still play a crucial role in this approach, since they manage the conversion regarding time.

4 Example

To demonstrate the feasibility of the approach, we implement a transceiver system using the TUV building blocks library, and show how

- The executable specification can be refined by adding non-ideal effects from analog subsystem to the behavior, and
- How we can attach a HW/SW system running on a DSP with two shared memories to the transceiver subsystems.

In the following, we won't describe real system synthesis. Instead, we will show that there is a seamless path from executable specification to architecture exploration and finally to the start of circuit design in analog and digital domain. As example, we use an OFDM transceiver that is modeled using the TU Vienna Building Block library for communication system design (source code: `www.systemc-ams.org`).

Executable specification – For the executable specification, most DSP tasks of the baseband processing are available in the library, modeled using the TDF model of computation. Furthermore, the executable specification includes some “analog blocks” such as quadrature mixer and LNA. These blocks are obviously analog, because they just modulate the information on specified physical quantities (modulated carrier). Apart from the pure transceiver, we assume that there is a protocol stack that handles media access (MAC layer) and other issues above this layer. Most of these tasks involve complex control flow and communication with application software. They are described best using the KPN model of computation. Figure 8 shows an overview.

Architecture Mapping and Architecture Exploration – The OFDM transceiver system offers directly a good partitioning (like most transceiver systems). Following the rules given in Sect. 3, we can map the executable

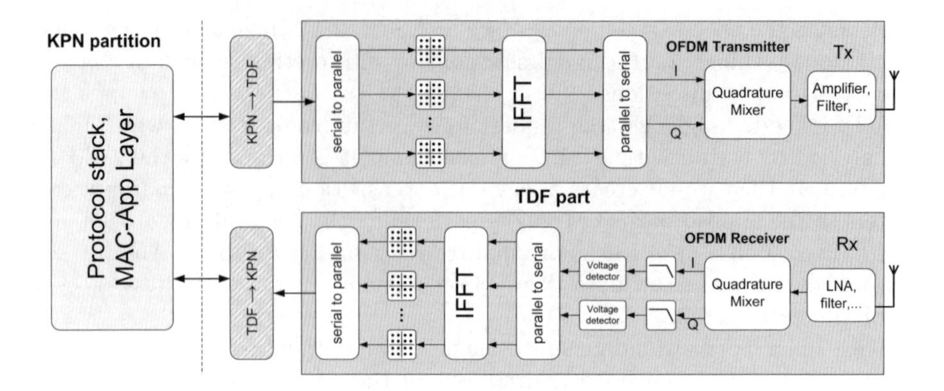

Fig. 8. Executable specification of an OFDM transceiver that combines KPN and TDF with embedded continuous-time models (filter, LNA)

specification as follows to architecture level processors, describing an initial mapping:

- Mixer, LNA, and filters must be implemented in analog due to excessively small signals and high frequencies and very low voltages (rule 2; signal conditioning)
- The baseband processing (FFT, IFFT, etc.) is described using TLM, can be implemented in SW, digital HW, or analog. Rule 3 recommends digital implementation.
- The protocol stack is modeled using KPN and has a quite complex control flow which demands implementation using SW.

Architectures to be investigated by architecture explorations are due to different mapping of the TDF part behind the mixer to DSP processor, digital, or analog hardware. The mapping of the TDF processes to DSP or dedicated digital HW basically spans the whole space of "SW defined" vs. "SW controlled radio". An interesting tradeoff specifically possible with HW/SW/Analog Co-Design is also the placing of.

Critical issues during architecture exploration are:

- Impact of jitter, noise, distortions, accuracy of A/D conversion on the dependability of communication.
- Performance of DSP subsystem under real operation conditions (i.e. shared memory architecture).

The DSP uses TDF models of an OFDM transmitter and of an OFDM receiver to receive and transmit data packets. These data packets contain WRITE and READ request, using a simple protocol, and refer to write and read transactions on the memory modules in the TLM part of the system. The DSP sends the requests to the memory via the bus, and returns the results (either the data for a READ request or an acknowledge for the WRITE request, or a fail in both cases). The TDF testbench provides the test data, and reads the results. The

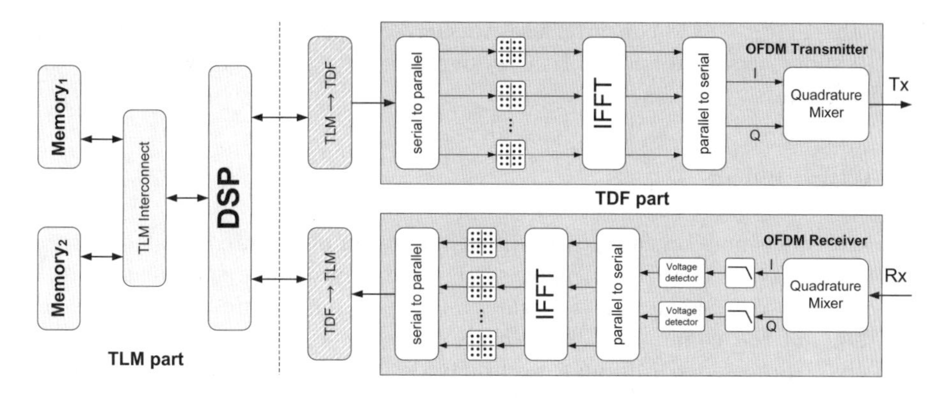

Fig. 9. Model of OFDM transceiver for architecture exploration: Analog hardware (mixer, filter, frontend, ADC, DAC) and DSP hardware (FFT, IFFT, serializer, deserializer, etc.) are modeled using TDF model of computation; DSP for protocol stack is modeled using TLM style communication

carrier frequencies of the incoming and the outgoing data are different (10 MHz and 5 MHz, respectively).

The DSP basically constitutes the wrapper for the TLM-TDF conversion mentioned above. Since the requests have different lengths, the incoming TDF data is stored in a buffer, and it is checked whether a complete request has been received yet. If so, an appropriate transaction is created and sent to the TLM-Interconnect. Regarding the memory responses, the DSP can simply mutate the transactions (e.g. add the protocol data to the data of a READ request), and pass it to the OFDM transmitter via the TLM to TDF converter. Figure 9 shows an overview of the example system used for architecture exploration. The TLM to TDF converter used here is the one described at the beginning of the Section,

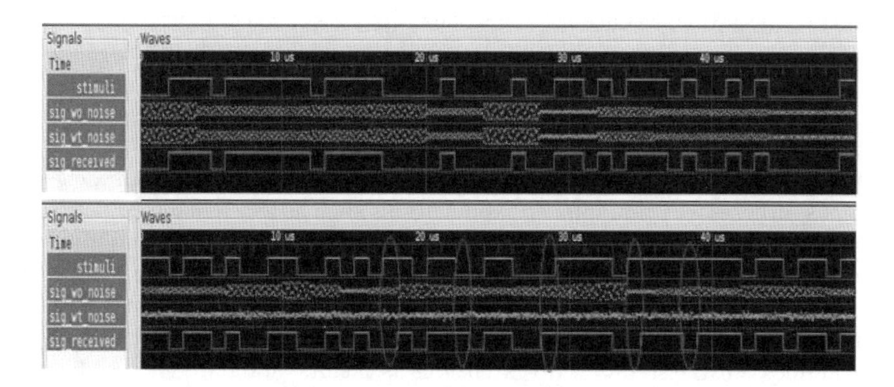

Fig. 10. Example screenshots of traces of the above model, one with an ideal channel without noise and attenuation (above), and one with noise, attenuation and resulting bit errors (below)

while the TDF to TLM converter uses the active approach described above. Figure 10 shows example traces of the system without noise and with noise and resulting errors.

5 Conclusion

We have described key ingredients that might enable future tools to achieve real Co-Design of HW/SW/Analog systems: An executable specification by a combination of TDF and KPN is the first step. Architecture mapping assigns partitions of the combined TDF / KPN processes to either analog or digital (HW/SW) processors. Although we briefly described some known simple rules and techniques for estimation of quantitative parameters that might assist automatic partitioning, partitioning will still require an overall system simulation. Only such a simulation is able to estimate most - sometimes very complex - properties of different architectures. For overall system simulation, we have discussed issues in co-simulation of TDF (=analog) and TLM (=HW/SW) systems that allows to raise the level of abstraction and to increase performance of simulation.

It seems that most ingredients for automatic synthesis are available. However, the devil is in the details: Properties of analog subsystems and the overall system cannot be estimated easily. Therefore, in the near future, system partitioning will still be based on experiences. One reason for this is that for estimation of HW/SW systems, complex functional units or even whole platforms can be assumed as pre-designed components with fixed properties, usually area, delay, power consumption. This is not the case for analog components: Analog design goes rather bottom up; therefore the output of HW/SW/Analog Co-Design is rather a starting point for a design, but for which properties (area, power consumption, and accuracy) are not known in advance.

Another major drawback are the have to be estimated for system synthesis including analog/RF systems are, in addition to area, delay and power consumption the properties accuracy and flexibility resp. re-configurability. Unfortunately, accuracy is very difficult to predict at higher level of abstraction if analog circuits are involved. Furthermore, flexibility and re-configurability are currently only understood as properties of an implementation. Means to specify this property as integral part of requirements are still missing.

Nevertheless, HW/SW/Analog Co-Design and thereby integral approaches for system syntheses will gain impact: Applications such as wireless sensor networks, energy management in future smart grids, or the electric vehicle underline the need for more comprehensive design methods that also take care of analog/RF or analog power components as vital parts of IT systems. However, like - and even more - than in HW/SW Co-Design modeling and simulation will play a central role. With SystemC AMS extensions as the first standard for specification of HW/SW/Analog Systems at system level, the most important step towards HW/SW/Co-Design has been done and is being adopted by industry.

References

1. Christen, E., Bakalar, K.: VHDL-AMS – A hardware description language for analog and mixed-signal applications. IEEE Transactions on Circuits and Systems-II: Analog and Digital Signal Processing 46(10) (1999)
2. Standard: IEEE: Verilog. Verilog 1364–1995 (1995)
3. Grimm, C., Barnasconi, M., Vachoux, A., Einwich, K.: An Introduction to Modeling Embedded Analog/Digital Systems using SystemC AMS extensions. In: OSCI SystemC AMS distribution (2008), Available on `www.systemc.org`
4. Huss, S.: Analog circuit synthesis: a search for the Holy Grail? In: Proceedings International Symposium on Circuits and Systems (ISCAS 2006) (2006)
5. Grimm, C., Waldschmidt, K.: Repartitioning and technology mapping of electronic hybrid systems. In: Design, Automation and Test in Europe (DATE 1998) (1998)
6. Oehler, P., Grimm, C., Waldschmidt, K.: A methodology for system-level synthesis of mixed-signal applications. IEEE Transactions on VLSI Systems 2002 (2002)
7. Lopez, J., Domenech, G., Ruiz, R., Kazmierski, T.: Automated high level synthesis of hardware building blocks present in ART-based neural networks, from VHDL-AMS descriptions. In: IEEE International Symposium on Circuits and Systems 2002 (2002)
8. Zeng, K., Huss, S.: Structure Synthesis of Analog and Mixed-Signal Circuits using Partition Techniques. In: Proceedings 7th International Symposium on Quality of Electronic Design (ISQED 2006) (2006)
9. Zeng, K., Huss, S.: RAMS: a VHDL-AMS code refactoring tool supporting high level analog synthesis. In: IEEE Computer Society Annual Symposium on VLSI 2006 (2005)
10. Zeng, K., Huss, S.: Architecture refinements by code refactoring of behavioral VHDL-AMS models. In: IEEE International Symposium on Circuits and Systems 2006 (2006)
11. Grimm, C.: Modeling and Refinement of Mixed-Signal Systems with SystemC. SystemC: Language and Applications (2003)
12. Klaus, S., Huss, S., Trautmann, T.: Automatic Generation of Scheduled SystemC Models of Embedded Systems from Extended Task Graphs. In: Forum on Design Languages 2002 (FDL 2002) (2002)
13. Kahn, G.: The semantics of a simple language for parallel programming. In: Rosenfeld, J.L. (ed.) Information Processing 1974, IFIP Congress (1974)
14. Lee, E., Park, T.: Dataflow Process Networks. Proceedings of the IEEE (1995)
15. Eker, J., Janneck, J., Lee, E., Liu, J., Liu, X., Ludvig, J., Neuendorffer, S., Sachs, S., Xiong, Y.: Taming Heterogeneity – the Ptolemy Approach. Proceedings of the IEEE 91 (2003)
16. Lee, E., Sangiovanni-Vincentelli, A.: A Framework for Comparing Models of Computation. IEEE Transactions on Computer-Aided Design of Integrated Circuits and Systems (1998)
17. Jantsch, A.: Modeling Embedded Systems and SoCs. Morgan Kaufman Publishers, San Francisco (2004)
18. Cai, L., Gajski, D.: Transaction level modeling in system level design. Technical Report 03-10. Center for Embedded Computer Systems, University of California (2003)
19. Aynsley, J.: OSCI TLM-2.0 Language Reference Manual. Open SystemC Initiative (2009)

20. Haubelt, C., Falk, J., Keinert, J., Schlicher, T., Streubühr, M., Deyhle, A., Hadert, A., Teich, J.: A SystemC-based design methodology for digital signal processing systems. EURASIP Journal on Embedded Systems (2007)
21. Ou, J., Farooq, M., Haase, J., Grimm, C.: A Formal Model for Specification and Optimization of Flexible Communication Systems. In: Proceedings NASA/ESA Conference on Adaptive Hardware and Systems (AHS 2010) (2010)
22. Parks, T.M.: Bounded Scheduling of Process Networks. Technical Report UCB/ERL-95-105. EECS Department, University of California (1995)
23. Rosinger, S., Helms, D., Nebel, W.: RTL power modeling and estimation of sleep transistor based power gating. In: 23th International Conference on Architecture of Computing Systems (ARCS 2010) (2010)
24. Henkel, J., Ernst, R.: High-Level Estimation Techniques for Usage in Hardware/Software Co-Design. In: Proceedings Asian Pacific Design Automation Conference (ASP-DAC 1998) (1998)
25. Moreno, J., Haase, J., Grimm, C.: Energy Consumption Estimation and Profiling in Wireless Sensor Networks. In: 23th International Conference on Architecture of Computing Systems, ARCS 2010 (2010)
26. Wolf, F., Ernst, R.: Execution cost interval refinement in static software analysis. Journal of Systems Architecture 47(3-4) (2001)
27. Haase, J., Moreno, J., Grimm, C.: High Level Energy Consumption Estimation and Profiling for Optimizing Wireless Sensor Networks. In: 8th IEEE International Conference on Industrial Informatics (INDIN 2010) (2010)
28. Bachmann, W., Huss, S.: Efficient algorithms for multilevel power estimation of VLSI circuits. IEEE Transactions on VLSI Systems 13 (2005)
29. Adhikari, S., Grimm, C.: Modeling Switched Capacitor Sigma Delta Modulator Nonidealities in SystemC-AMS. In: Forum on Specification and Design Languages (FDL 2010) (2010)
30. Damm, M., Grimm, C., Haase, J., Herrholz, A., Nebel, W.: Connecting SystemC-AMS models with OSCI TLM 2.0 models using temporal decoupling. In: Forum on Specification and Design Languages (FDL 2008) (2008)
31. SystemC AMS Users Guide, OSCI (2010), `www.systemc.org`
32. Alur, R., Courcoubetis, C., Halbwachs, N., Henzinger, T.A., Ho, P.H., Nicollin, X., Olivero, A., Sifakis, J., Yovine, S.: The Algorithmic Analysis of Hybrid Systems. Theoret-ical Computer Science 138(1), 3–34 (1995)
33. Grimm, C., Heupke, W., Waldschmidt, W.: Analysis of Mixed-Signal Systems with Affine Arithmetic. IEEE Transactions on Computer Aided Design of Circuits and Systems 24(1), 118–120 (2005)

A Flexible Hierarchical Approach for Controlling the System-Level Design Complexity of Embedded Systems

Stephan Klaus

Technische Universität Darmstadt
Department of Computer Science
Integrated Circuits and Systems Lab
Hochschulstraße 10
64289 Darmstadt, Germany

Abstract. This paper summarizes results of the PhD theses on system-level design of embedded systems at the Institute Integrated Circuits and Systems Lab under the survey of Prof. Dr.-Ing. Sorin Huss. A straightforward identification of suited system-level implementations of distributed embedded systems is increasingly restricted by the complexity of the solution space and the size of modern systems. Therefore, concepts are mandatory, which are able to control the design complexity and assist the reuse of components. A hierarchical, task-based design approach and two algorithms are developed, which allow to derive dynamically partial specification models for design space exploration on different levels of detail as well as task descriptions for IP encapsulation So, the descriptive complexity of specifications is considerably reduced and the execution time of system-level synthesis algorithm can be adopted to the current requirements of the designer. The task behavior is captured by Input/Output Relations, which represent a very general and powerful means of encapsulating internal implementation details and of describing data as well as control-flow information on different levels of detail. The necessity of these concepts are demonstrated by means of an MP3 decoder application example.

1 Introduction

Since the beginning of computer-aided hardware synthesis in the early 70's, the progress can be characterized by a steady shift of the design entry towards higher levels of abstraction. At present, researchers focus on system-level design, which addresses the synthesis of complete systems based on coarse-grained specifications. This process can be detailed as the selection of the necessary resources (allocation) and the mapping of functional units onto the selected architecture in space (binding) and time (scheduling), whereby real-time and other constraints must be met. In particular, the non-functional requirements of embedded systems such as timing, cost, power consumption, etc., are extremely relevant for their later commercial success. This optimization process is restricted both by

A. Biedermann and H. Gregor Molter (Eds.): Secure Embedded Systems, LNEE 78, pp. 25–42.
springerlink.com

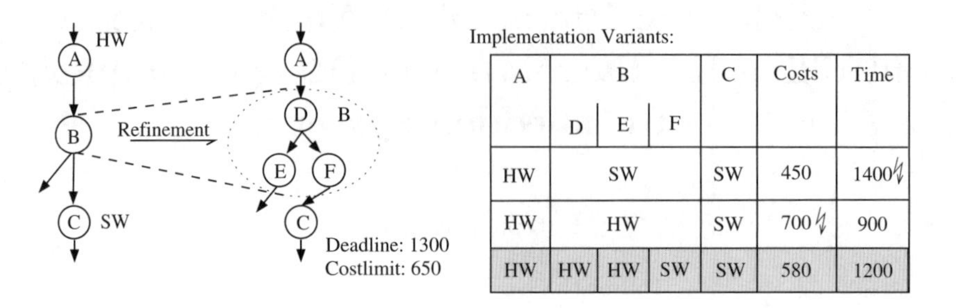

Implementation Variants:

A	B			C	Costs	Time
	D	E	F			
HW	SW			SW	450	1400↯
HW	HW			SW	700↯	900
HW	HW	HW	SW	SW	580	1200

Fig. 1. Implementation Variants considering different Levels of Detail

increasing complexity of systems and by shorter time to market, whereby the complexity in task-based specification models and therewith the execution of synthesis algorithms grows exponential with the number of tasks. A tasks represents a single, quite simple part of an algorithm of the complete specification, e.g. the Huffman decoding for an MP3 decoder. Even medium scaled embedded systems are so complex, that flat representations become too difficult for humans to handle and to comprehend. In the same manner as flat representations become too complex, the execution time of synthesis and design space exploration tools grows unacceptably, e.g., the simulation of an MP3 decoder takes $901s$ at instruction set level, while a transaction level simulation takes only $41s$. Due to this fact, the design flow including all necessary implementation steps should be accomplished on a current adequate level of detail only. But the drawback of too abstract specifications is a loss of resulting quality, since the chosen level of detail directly influences the optimization potential. Figure 1 illustrates this problem by means of a hardware/software partitioning problem of task B. As it can be seen, the cost and latency deadlines can not be fulfilled if task B is completely implemented in software (first row) or hardware (second row). Instead of implementing task B completely in hardware or software, a mixed implementation is chosen, whereby task B is refined into a more detailed model consisting of the three tasks D, E and, F. So, task D and E are implemented in hardware, while F is mapped to software.

Such local refinements lead to the idea of static hierarchical synthesis algorithms. Hierarchical algorithms work recursively on the static specification hierarchy, whereat each module with its subtasks is optimized individually and the results are composed on the next higher level. This reduces the design complexity since smaller models have to be synthesized, but optimization potential is dropped away. The loss of optimization potential between static hierarchical synthesis algorithm and a global approach is depicted in Figure 2. The composition of the individual results on the next higher level does not allow to schedule task E in parallel to task $C2$, which is possible applying the flat synthesis process. Nevertheless, the complexity in the hierarchical case is considerably reduced, since in task-based specification models the complexity grows exponential with the number of tasks.

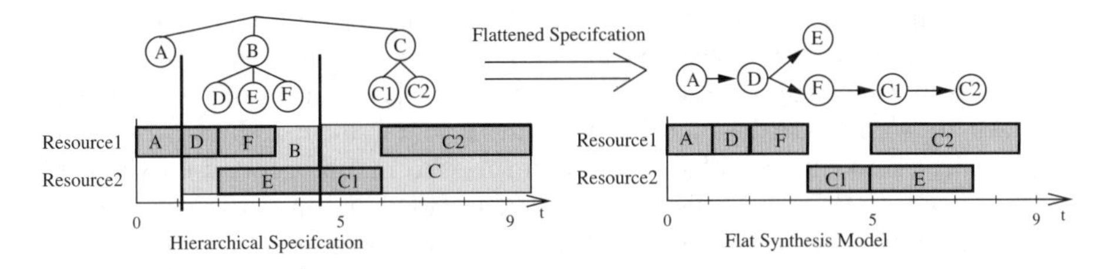

Fig. 2. Local versus Global Optimization

It can be summarized that the level of detail of a specification is the crucial point in terms of resulting quality and of execution time of the system-level synthesis steps. A partial refinement of specifications may be mandatory to fulfill all design requirements. On the other hand some parts of the specification may be merged into one module to reduce the complexity and therewith the execution time of the system-level synthesis. This means that a hierarchical specification model has to provide dynamically suited representations for any subpart of the specification and at any level of detail. Such a multi-level representation should be able to capture the external behavior of subtasks and to represent data and control-flow on different levels of detail, whereby internal implementation details should be hidden away. Thus, a reduction of the descriptive complexity of embedded systems and both definition and reuse of IP cores become possible.

The next section discusses some other work, related to the presented methods. Section 3 outlines the proposed computational specification model and its means of system behavior description. Algorithms for generating dynamically flat model views and system behaviors on different levels of abstraction are detailed next. Then, Section 5 summarizes the results produced by means of a real-life application example of an MP3 decoder. Finally, a conclusion is given and some open questions and future work are discussed.

2 Related Work

Designers of embedded systems can choose from many rather different specification models. A good survey and a detailed description of appropriate specification models can be found in [6]. According to [10] these models can be grouped into five different classes: State-oriented, activity-oriented, structure-oriented, data-oriented, and heterogeneous models. In contrast to state-oriented models like Statecharts [11] or Esterel [2], which support hierarchy, most task-based specifications are flat models. In the area of high-level synthesis hierarchical models were explored in the past. An overview can be found in [17]. Hierarchical system-level synthesis based on task graph specifications is addressed in [19,12,7]. [12] facilitate a classical hierarchical bottom-up design, whereby the determined

Pareto optimal solutions are combined on higher levels of abstraction. A simulation based approach is presented in [19]: The level of detail is increased each time the solution space is reduced by system-level design decisions. COHRA [7] proposes both a hierarchical specification and a hierarchical architecture model. A specialized co-synthesis algorithm performs the mapping between these hierarchical models. But all these approaches are working on static specification hierarchies only and need specialized algorithms. Therefore, these approaches are not so flexible and optimization just can take place internal to the fixed functional partitioning as demonstrated in Figure 2 and therefore a loss of optimization flexibility is in general to be expected.

Beside pure data-flow applications captured by task graph models as mentioned above, embedded systems mostly include control-flow parts, which must be definitely taken into account for a complete and correct specification. Therefore, a lot of research was done extending task graphs by control-flow information. The CoDesign Model (CDM) [4] and the System Property Intervals (SPI) model [9] [21] are cyclic graph models, which capture the behavior by relating input edges to output edges. Thereby each task can have more than one of such relations to model different control-depending system reactions. In addition, the SPI model supports intervals for the timing and the amount of data. Conditional Process Graphs (CPG) [8] and extended Task Graphs (eTG) [16] model control-flow by boolean conditions, which are assigned to data-flow edges. In addition, eTG support conditional behavior on input edges too. But all of these models do not allow for a hierarchical specification, which is mandatory for a system-level design of large systems.

3 Computational Model

The proposed specification method denoted as hierarchical CoDesign Model (hCDM) is intended as a means for the description of IP core encapsulation and for system-level synthesis on different levels of detail. The model is a specialization and hierarchical extension of [4] and is very well suited for data-flow dominated systems, which contain control-flow behavior, too. In addition to the hierarchical extension of the CDM, the data cycles in the original model is removed from all definitions in order to allow an efficient system synthesis. The cyclic property and the extended behavior description of the CDM lead to Turing complexity of the model [5] and therewith the complexity for scheduling problem is NEXPTIME [13], which is not accabtable for system-level synthesis. Also without data cycles, periodic behavior and bounded loops can be modeled in hCDM. So, without loosing too much modeling power, the synthesis complexity is drastically reduced. For that purpose also the behavior description and the definitions of the conditions must be adopted.

The novel dynamic hierarchical decomposition property and the uniform means of behavior description are its main advantages compared to state of art system-level computational and specification models for embedded systems.

3.1 Hierarchical Specification Method

A hCDM is a directed acyclic graph consisting of computational tasks, inputs, outputs, and dependencies between the tasks. Additionally, each task may be defined as a separate hCDM thus introducing recursively a hierarchical structure.

Definition 1. *(hCDM Graph):*
The hierarchical CoDesign Model consists of a directed acyclic graph $G = (V, E)$ *where the nodes are composed of inputs, tasks, and outputs* $V = I \cup T \cup O$*. An arc* $e \in E$ *is a 2-tuple* (v_i, v_j), $v_i, v_j \in V$*. In this model*

- I *denotes the set of inputs:* $\forall i \in I. \neg\exists(v, i) \in E.v \in V$
- O *denotes the set of outputs:*$\forall o \in O. \neg\exists(o, v) \in E.v \in V$
- T *is the set of computational tasks:* $\forall t \in T.\exists(t, v_1) \in E \wedge \exists(v_2, t) \in E.v_1 v_2 \in V$.

Each task $t \in T$ *in a hCDM* G *can be defined recursively as a separate hCDM denoted as* $G' = (V', E')$*. Thereby each input edge of* t $(v, t) \in E, v \in V$ *is assigned to an input* $i' \in I'$ *and each outgoing edge* $(t, v) \in E, v \in V$ *is assigned to an output* $o' \in O'$ *of* G'*.*

Figure 3 illustrates a hierarchical specification by means of a simple example, whereas the top-level module T is subdivided into three tasks A, B, and C. The assignment of inputs/outputs to edges of the parent hCDM introduces hierarchical edges and thus a hierarchical data-flow. These edges are visualized as the links between the levels n, $n+1$ and $n+2$.

In contrast to this hierarchical problem specification, a current model view defines a sub-area of interest. All tasks, which are below a current level of detail in the specification hierarchy, are ignored for the current model view and just

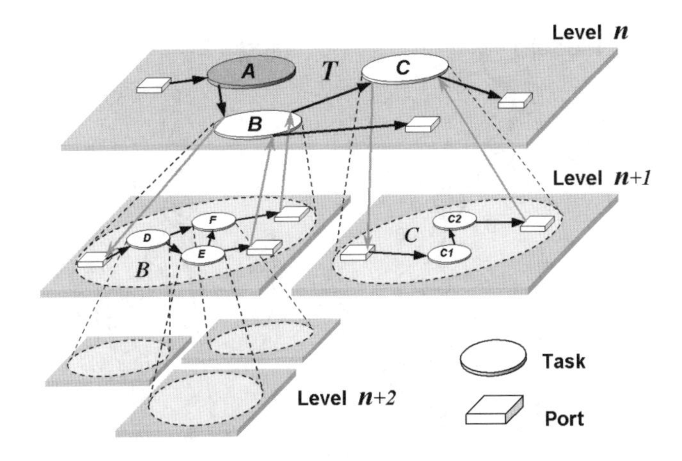

Fig. 3. Hierarchical Specification Example

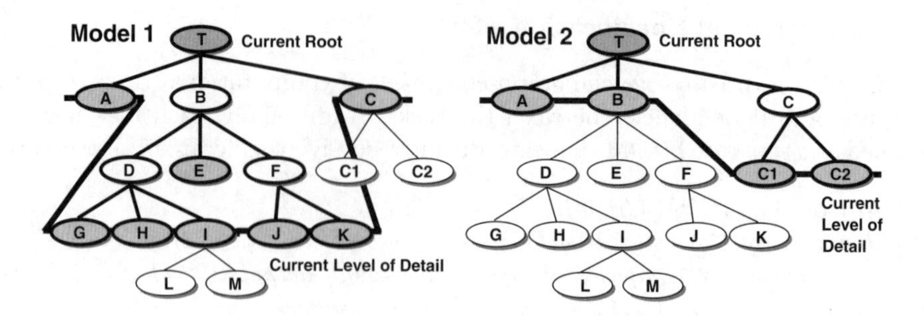

Fig. 4. Current Model View Examples

the subtree starting at a current root is taken into account. Note that the tasks of the current level of detail are not necessarily leaves of the original hCDM specification.

Definition 2. *(Current Model View):*
The current model view of a hCDM G is defined by a 2-tuple $A = (r, L)$ where

- *$r \in T$ denotes the current root task and*
- *$L \subseteq T$ the current level of detail.*

Figure 4 illustrates the definition of the current model view based on the complete hierarchical view of the example in Figure 3. Two different model views are illustrated: Task T represents the root of both submodules and A, G, H, I, J, K, C the current level of detail of the first model view. The second model encapsulates task B on a higher level of abstraction leaving out the detailed subdivision into task D up to task K. Additionally, task C is refined into $C1$ and $C2$. The

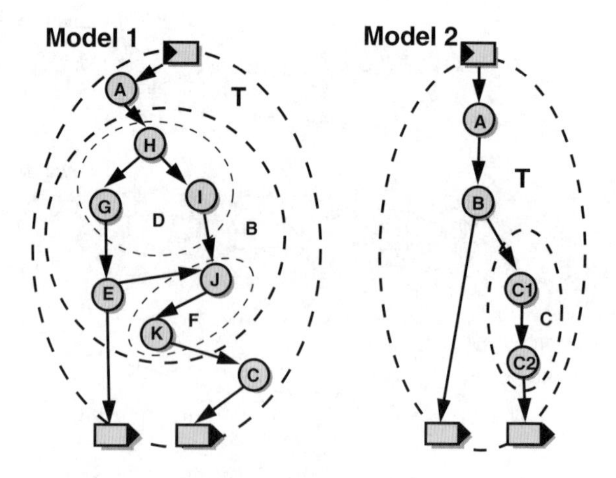

Fig. 5. Partially flattened hCDM Models

sub-area defined by a current model view can be transformed into a flat task graph specification. So, standard synthesis algorithm for partitioning, allocation, binding and, scheduling can be applied in a straight forward manner. Figure 5 depicts the transformed flat hCDMs for both models of Figure 4. The dashed lines denote the hierarchy of the original model and the arrows visualize the data-flow.

In Section 4 algorithms are proposed to derive automatically flat current model views according to user requests. So, it is possible to manipulate dynamically the hierarchical structure in a systematic way, which allows to adopt the level of detail to the current necessary level in a flexible way. Thus, the complexity of the system-level synthesis steps can be easily reduced to a necessary minimum with respect to the required accuracy.

3.2 Dynamic System Behavior

A hCDM is intended to jointly represent data and control-flow information. Transformative applications like digital signal processing in multimedia applications, e.g., MP3 decoder or encoder, are dominated by data-flow operations, but include also a few control-flow decisions. Both data- and control-flow information of tasks should be captured on different levels of detail by means of one and the same mechanism. In case of modeling the control-flow, a more sophisticated behavior description than a simple Petri-net [20] activation rule is mandatory, which is frequently exploited in task graph specifications.

Figure 6 illustrates code fragments, which cause possibly different control-flow behaviors. The output of a task, which contains such code fragments, may differ according to the value of *type1* in the switch statement, or to a in the if statement, respectively. In order to address this property Definition 3 introduces behavior classes to denote possibly different control-flow effects in a hCDM. A behavior class consists of an identifier and of a set of different behaviors according to possible control-flow decisions.

Definition 3. *(Behavior Class):*
A behavior class b is a 2-tuple $b = (n, W)$, where n is a unique identifier and W a set of at least two different behaviors. B_G is the set of all possible behavior classes in a hCDM G.

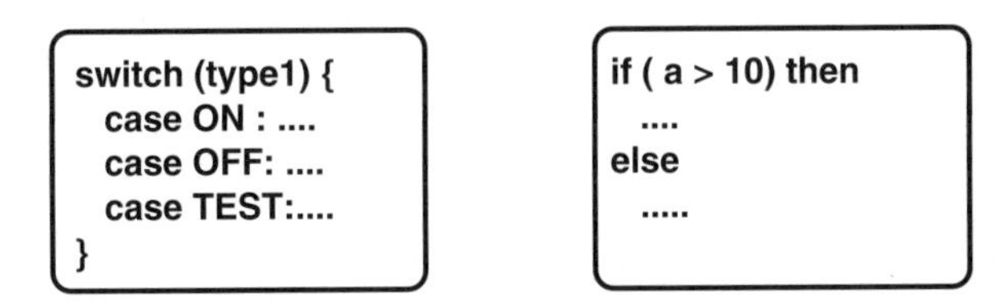

Fig. 6. Behavior Class Examples

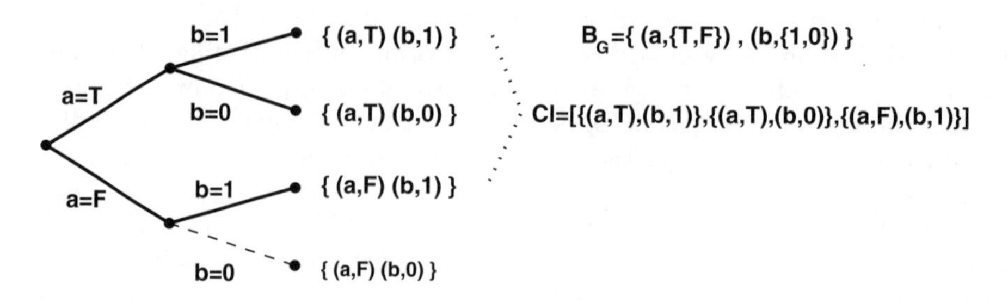

Fig. 7. Conditionlist Example

The Behavior Classes for the examples in Figure 6 result in:

$$b_{left} = (type1, \{ON, OFF, TEST\}) \text{ and } b_{right} = (a, \{T, F\}).$$

Since a behavior class defines the set of all possible externally observable system reactions, a Condition restricts this set to one current valid behavior. During execution of a system a current control-flow reaction is selected from the set of all possible behaviors. Therefore, a condition embodies the control-flow decisions taken.

Definition 4. *(Condition, Conditionlist):*
A condition c is a 2-tuple $c = (n, w)$, where $w \in W$ with $(n, W) \in B_G$.
A sequence of conditions is a Condition-Sequence Z.
A set of Condition-Sequences is a Conditionlist C.

A Condition-Sequence denotes conditions, which are executed in sequel. Figure 7 presents a decision tree consisting of two Behavior Classes, each featuring two different behaviors. The nodes of the tree correspond to control-flow decisions taken during the execution of a system. The two Behavior Classes result in 2^2 different system reactions, whereby the resulting Condition-Sequences are attributed to the leafs of the decision tree in Figure 7. A Conditionlist subsumes different Condition-Sequences to capture identical system output for a class of Condition-Sequences. Cl in Figure 7 includes all Condition-Sequences except $\{(a, f)(b, 0)\}$. So, Cl describes all executions when $a = T$ or $b = 1$. In order to map and to transfer the behavior to different levels of abstraction, some operators are to be introduced for the manipulation of Conditionlists. Especially, the union and the intersection of these lists, performed by means of three auxiliary functions, are to be provided.

Definition 5. *(Operators):*
Two operators on Conditionlists denoted as $\oplus$, $\ominus$ and three auxiliary functions Names, All, Expand are defined as follows:

– **Names**: $C \rightarrow \{n_1, .., n_i\}$
  $Names(C) = \{n | (n, v) \in C\}$

– **All**:$\{n_1, .., n_i\} \to C$
$$All(\{n_1, ..n_i\}) = \bigcup_{\substack{\forall s_1 \in W(n_1) \\ \dots \\ \forall s_i \in W(n_i)}} \{(n_1, s_1), .., (n_i, s_i)\}$$
– **Expand** $C \times \{n_1, .., n_i\} \to C$
$$Expand(C, \{n_1, .., n_i\}) = \bigcup_{\substack{\forall Z \in C \\ \forall s_1 \in W(n'_1) \\ \dots \\ \forall s_i \in W(n'_i)}} \{Z \cup \{(n_1, s_1), .., (n_i, s_i)\}\}$$
with $\{n'_1, ..n'_i\} = \{n_1, ..n_i\} \setminus Names(Z)$
– $\ominus : C \times C \to C$
$$C_1 \ominus C_2 = Expand(C_1, Names(C_1 \cup C_2)) \cap Expand(C_2, Names(C_1 \cup C_2))$$
– $\oplus : C \times C \to C$
$$C_1 \oplus C_2 = Expand(C_1, Names(C_1 \cup C_2)) \cup Expand(C_2, Names(C_1 \cup C_2))$$

The operator $Names$ determines the set of all identifiers in a Conditionlist. According to a set of Behavior classes *All* calculates a Conditionlist, which contains all possible Condition-Sequences. *Expand* takes a Conditionlist and a set of Behavior Classes as an input and produces an identical Conditionlist extending all Condition-Sequences of the input Conditionlist with the additional Behavior Classes. This allows the application of simple set operations for $\oplus$ and $\ominus$. Accordingly, $\ominus$ defines the intersection of two Conditionlists. The resulting Conditionlist contains the Condition-Sequences, which are contained in C_1 and C_2. In contrast, $\oplus$ defines the union of two Conditionlists. The resulting Conditionlists is valid, if Conditionlist C_1 or C_2 is valid.

3.3 Task Behavior

The external behavior of tasks should be captured by a uniform means of description. For example, the MP3 decoder task *Stereo* in Figure 8 produces outputs on both outgoing edges, in case that the input stream is stereo coded. Otherwise, if the sound is mono encoded, only one output will be activated. Input/Output (IO)Relations detail this external system reactions. The input set defines the edges, which must present data to activate a task. Upon activation new data is then generated on all edges specified by the output set. The IO Relation becomes valid under the referenced Conditionlist only.

Definition 6. *(Input/Output Relation):*
For every $t \in T$ *a set of Input/Output Relations is assigned. An Input/Output Relation is a 3-tuple* $< in|out, C >$. *where:*

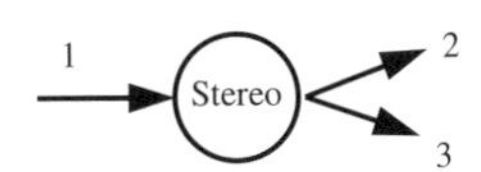

Fig. 8. Example for External Task Behavior

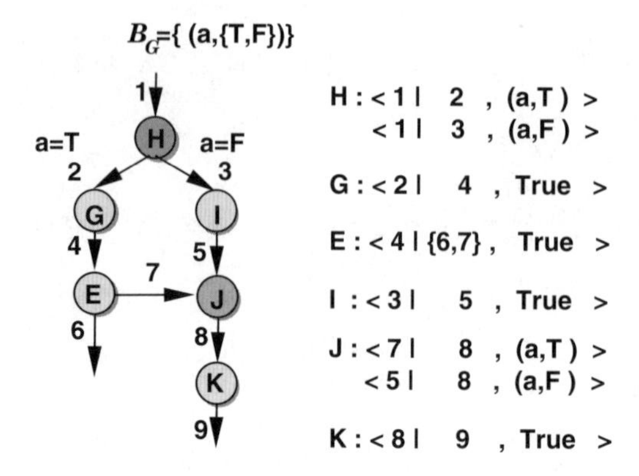

Fig. 9. Input/Output Relations

- *$in \subseteq \{(v,t)|(v,t) \in E \wedge v \in V\}$: subset of the input edges of t*
- *$out \subseteq \{(t,v)|(t,v) \in E \wedge v \in V\}$: subset of the output edges of t*
- *C denotes the Conditionlist.*

The IO Relations for the simple example of Figure 8 results in: $< 1|2, (t, mono) >$ and $< 1|\{2,3\}, (t, stereo) >$. Figure 9 illustrates another example for the definition of IO Relations by means of a small task graph example. Later on, the outlined behavior will be subsumed to a high-level description. Figure 9 details one Behavior Class named a, which contains two different behaviors. Task H splits the control-flow according to Condition a, while task J merges the different paths. The output on edge 6 occurs only if the value of Condition a is T. The other tasks need data on all their input edges and upon activation new data on all their output edges is generated. Conditional Process Graphs [8] and extended Task Graph [16] specifications can be easily transformed into this more general specification model. The task behavior of SDF [18] specifications can be modeled by the proposed IO Relations too.

In order to perform a design space exploration an appropriate and well-defined execution semantic is mandatory. The activation of tasks takes place according to the activation rule outlined in Definition 7 considering the following basic principles: The execution of any task is non-preemptive and inter-task communication may only take place either at the start or upon termination of a computation.

Definition 7. *(Activation Rule):*
A task t in a hCDM G can be activated by a set of data $M \subset E$ under Conditionlist C_m, if an IO Relation $io =< in|out, C_{io} >$ exits in t, with:
$in \subseteq M$ and $C_m \ominus C_{io} \neq \emptyset$.

A task can be only activated if valid data on its specified input edges is available and the Conditionlists match the Conditionlist of the current execution. Upon activation, the resulting data in G is determined from an Update Rule as follows.

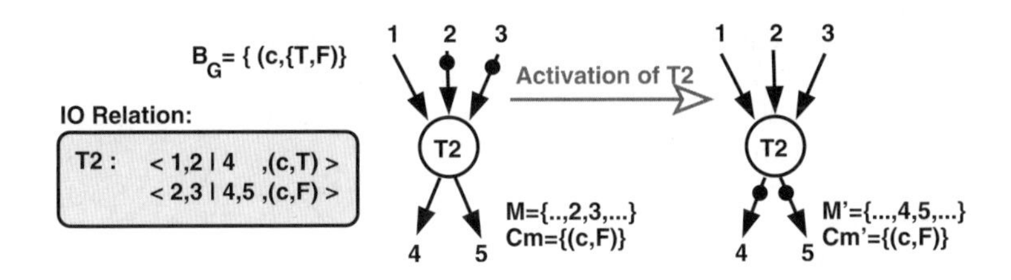

Fig. 10. Task Execution

Definition 8. *(Update Rule):*
Upon activation of task t in a hCDM G under IO Relation $io =< in|out, C_{io} >$ the set of data M and the Conditionlist of the current execution C_m is updated to M', C'_m according to the following properties:

- $M' = M \setminus in \cup out$
- $C'_m = C_m \ominus C_{io}$.

If an IO Relation is activated, its input data is removed and new output data on all edges denoted in the output set is generated. In case that a new condition is introduced by the activation of the IO Relation, then the Conditionlist of the current execution has to be adjusted accordingly.

Figure 10 illustrates a task denoted $T2$ and its IO Relations as an example. This task may be activated under Condition (c, T) consuming data from input edge 1 and 2. Alternatively, it may be activated under Condition (c, F) reading data from edge 2 and 3 generating new data on edge 4 and 5, respectively. As depicted by the filled circles in Figure 10 edges 2 and 3 contain data, so $T2$ is being activated under Condition (c, F). The set of edges, which contain data M is then changed as visible from the right part of Figure 10.

4 Generating Current Model Views

To overcome the limitations of a static hierarchical structure two algorithms are presented in the sequel aimed to the derivation of flat subgraphs at variable levels of detail. These algorithms are intended to overcome the limitations of static hierarchical structures. The algorithm of Figure 11 calculates the flat hCDM defined by a given current root, a current level of detail, and a complete hCDM. So, multiple views on different levels of detail can be generated dynamically. The static structures are traversed by this flexible method. Flat representations generated in this way allow for the application of standard algorithms for the subsequent design steps.

The ports of the root task become the inputs (loop line 1 in Figure 11) and outputs (loop in line 8) of the current model view. The set of tasks is established by traversing the hierarchy tree starting from the root of the current model view down either to a leaf or to a task in the current level of detail (while loop in line 4). This restricts the current task list to a well-defined cut tree, since L in

```
In: G: hCDM ; r: Current Root ; LoT: Current Level of Detail
OUT: aTasks: Current Task List; aInputs: Current Input Ports
     aOutputs: Current Output ports; aEdges: Current Data Dependencies
(1) for all  ( p ∈ r.ports ∧ p.type=IN)do
(2)     aInputs.insert(p) /* Insert Input */
(3) testTasks= r.GetSubTasks()
(4) while (testTasks.empty()=FALSE) do
(5)       t=testTasks.removeFirst()
(6)       if (t ∈ LoT ∧ t.GetSubTasks()=∅) then aTasks.insert(t) /* Insert Task */
(7)       else testTasks = testTasks ∪ t.GetSubTasks()
(8) for all (p ∈ r.ports ∧ p.type=OUT ) do
(9)       aOutputs.insert(p) /* Insert Output */
(10) for all ( n ∈ aTasks ∪ aInputs ) do
(11)       for all ( h ∈ n.GetHierachicalOutEdges() ) do
(12)            n′=h.findSink(t,aTasks ∪ aOutputs)
(13)            aEdges.insert(edge(n,n′))) /* Insert Data-Flow Edge */
```

Fig. 11. Algorithm: BuildCurrentModelView

Definition 2 may not necessarily define a proper cut tree. The data dependencies of the current model are determined by traversing the hierarchical data-flow edges starting from all current tasks and inputs up to a task or to an output of the current model view (loop in line 10). The complexity of the algorithm is $O(\#\text{tasks}*\#\text{edges})$ because of the loops in line 10 and 11.

After creating a partially flat representation of the hierarchical model, it is necessary to summarize the behavior of the subtasks. The algorithm of Figure 12 takes the IO Relations of these subtasks as input and derives a compact description of the envisaged high-level model.

This algorithm can be used in a bottom-up design style as well as in a top-down verification flow. In the first case the IO Relations are generated with respect to the behavior of the subtasks. In the second case it can be easily checked that the implementation of the subtasks is equivalent to the specification of the parent task. The algorithm takes a task as input and determines its IO Relations with respect to the behavior of its subtasks. This can be done recursively to bridge more than one level of abstraction. The generation process is illustrated in Figure 13 for the subgraph of Figure 9 thus producing an IP description for this subgraph.

At first, the set of all possible Conditionlists is determined considering the IO Relations (loop in line 2 of Figure 12) of all subtasks using the *All* operator (line 4). The example from Figrure 9 has two different Conditionlists depending on whether a is T or F, as it can be seen below the dashed lines in Figure 13. Then, the input and output sets for each Conditionlist (loop in line 5) are established considering all generated Conditionlists and the underlying connectivity. The input and output set are combined (line 12) if the union of the Conditionlists is not empty and the current IO Relation is linked to the newly build IO Relation (line 9). Internal data edges are removed (loop in line 11), visualized by the

```
In: t: Task
OUT: ioR: New Input/Output Relations
(1) N = ∅ ; ioR = ∅ ; ioR′ = ∅
(2) for all  ( io ∈ t.ioRelations )do   /* Determine Condition Lists*/
(3)      N=N ∪ Names( io.c)
(4) condSet = All(N)
(5) for all  (c ∈ condSet )do
(6)      in = ∅; out = ∅
(7)      while ( in or out changed ) do   /* Determine Input and Output Sets*/
(8)           for all  ( io ∈ t.ioRelations )do
(9)                if ( ( ( io.c ⊖ c) != ∅ ∧ (io.in has external port
                     ∨ in ∩ io.out ≠ ∅∨ out ∩ io.in ≠ ∅ ) ) ) then
(10)                 in = in ∪ io.in; out = out ∪ io.out
(11)          for all(e ∈ in ∧ e ∈ out ) do
(12)               in=in \ e ; out=out \ e
(13)     ioR′.insert (in,out,c)
(14) for all  (io ∈ ioR′ ) do   /* Compact IO Relation */
(15)      c=io.c
(16)      for all  (io′ ∈ ioR′ ) do
(17)           if (io.in = io′.in ∧ io.out = io′.out ) then
(18)                c= io.c ⊕ io′.c; ioR′ = ioR′\ io \ io′
(19)      ioR.insert(io.in,io.out,c)
```

Fig. 12. Algorithm: DeriveIORelFromSubTasks

canceled edges in step 3 of Figure 13. Finally, IO Relations (loop in line 14) are merged in case that the input and outputs sets are equal and all dispensable Conditionlists are removed (line 18). So, the complexity of the algorithm depends on the amount of behavior classes and IO Relations O(#Behavior Classes*#IO Relations*#Ports), because of the loops in line 5, 7 and 8. The resulting two IO Relations highlighted in Figure 13 represent the behavior of task B, which was previously defined by eight IO Relations and six subtasks as illustrated in Figure 9. Thus, the proposed approach to IO Relation encapsulation hides its internal implementation and defines the external view of the envisaged high-level model, i.e., an IP module. The proposed algorithms support designers of embedded systems in both generating flexible submodules on all desired levels of detail and in dealing with the dynamic control behavior of tasks on different levels of detail.

5 Results

This sections attempts the hard try to give an idea of the benefits and the application of the proposed specification model and its flexible hierarchical manipulations. Beside some generic, but meaningful experiments the system-level design of a real-time MPEG 1 Layer 3 (MP3) Decoder is presented. First of all it has to be mentioned that it is not in the focus of the paper and no easy answer

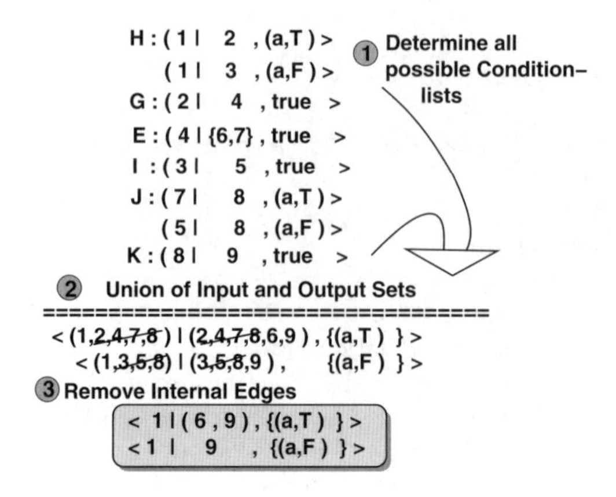

Fig. 13. Generation Process of IP Descriptions

can thus be given to the question of the right or optimal level of detail for the design process. However, some hints are given at the end of the section.

Figure 14 shows the execution time of a system-level synthesis algorithm depending on the number of tasks in the specification. The process of system-level synthesis or design space exploration is to determine suited implementation variants. For this purpose a genetic design space algorithm [14,15] is chosen. This heuristic is able to deal with different design goals, e.g., run-time and size. For this experiments ten individual runs were investigated for ten different specifications of each size. The specifications are generated by the tool Task Graph for Free (TGFF), which was developed by Robert Dick at Princeton University.

The execution time increases directly by the number of tasks. Thus it is obvious, that a specification should contain just the absolutely necessary amount of tasks. If the number of tasks can be reduced, then the execution time is also drastically reduced and design time can be saved. The proposed specification model and the novel algorithms are aimed to a fast and flexible adoption of the level of detail and thus to a reduction of the number of tasks for the system-level synthesis step.

The necessity of a flexible refining of a specification for the system-level design is demonstrated by means of an MP3 Decoder. The upper two levels of the hierarchical specification of the MP3 decoder, which will be implemented on top of an Atmel FPSLIC System-on-Chip (SoC) [1] is summarized in Figure 15. The specification is derived from a functional partitioning of the problem and the design process has to map this functional specification to the Field Programmable Gate Array (FPGA) or the AVR micro-controller of SoC. Additionally, the chosen genetic design space exploration calculates the complete system-level implementation including scheduling, allocation, and binding. At top-level the MP3 decoder functionality consists of reading data from a Multi Media Card, of decoding tasks, and of controlling the digital/analog converter. The decoding task is refined into 14 subtasks on the next level as shown in Figure 15. Here, the main tasks are the Huffman decoding (Huffman), the Inverse Modified Discrete

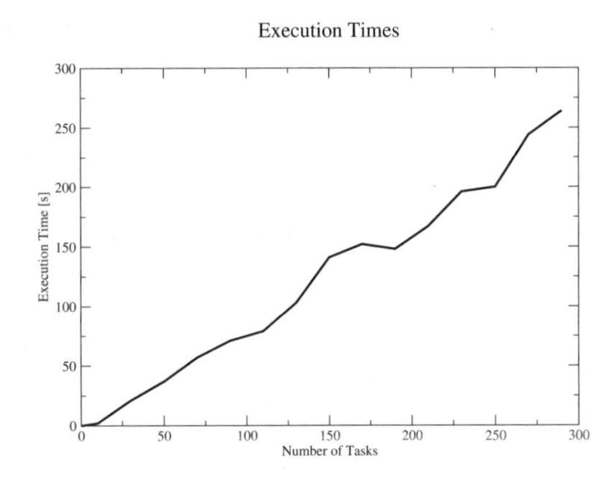

Fig. 14. Execution time of the system-level Synthesis for different levels of detail

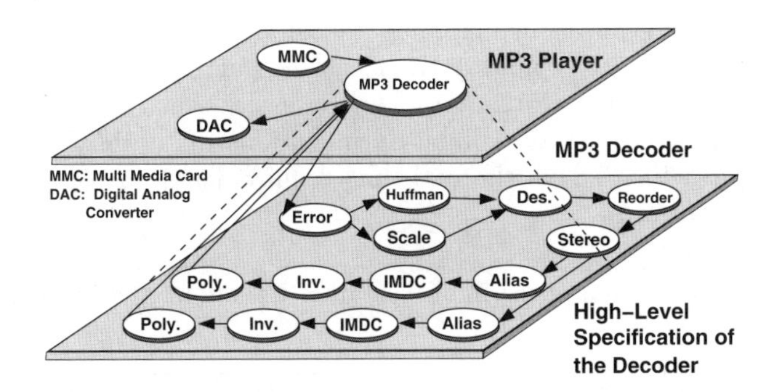

Fig. 15. Hierarchical Specification of an MP3 Decoder

Cosine Transform (IMDC), and the Synthesis Polyphase Filterbank (Poly). The most important non-functional specification of the decoder imposes that for real-time audio streaming at least 39 frames per second have to be decoded. The size of the FPGA and the rather small memory resources are the main limitations of this target SoC architecture, which have to be considered accordingly by the proposed design process.

Three different design approaches had to be investigated to finally receive a real-time implementation. First of all, a fast approach was chosen, which tries to map the high level specification of Figure 15 directly. The partitioning process assigns the data-flow dominated parts IMDCT, Inverse, and Polyphase Filter Bank to the FPGA, and the control-flow dominated parts to the AVR micro-controller. This design process is fast, since only 14 tasks have to be mapped, but the approach failed, however, either due to the limited memory or due to the FPGA resources. Therefore, in a second approach the preliminary implementation was improved by local optimizations of each task. This leads to a static hierarchical approach, where the hierarchical tasks are designed individually and combined on the next higher level. These separate modules for the tasks are generated by exploiting algorithm 11. The refined tasks contain between 45 and 442 subtasks. An optimization of the complete refined model is not applicable due to the execution time. The estimated execution time of an accurate simulation for the complete model is about $4,32 * 10^7$s. But this traditional hierarchical approach violates, again, the real-time specs. In a last step the flexible hierarchical approach was considered. Since the data-flow dominated parts on the FPGA are the bottleneck of the static hierarchical approach, the refined models of the IMDCT, Inverse, and Polyphase Filter Bank are merged into one submodule, which is then optimized in a partially global process. This submodule is generated by exploiting algorithm 12. The partially global submodule, which consists of about 600 subtasks, can now share resources and can finally be optimized in time as shown in Figure 2. This flexible approach leads to an acceptable real-time implementation of the decoder. A more detailed discussion of the design approach especially of the subsequent register transfer(RT) design for this application example can be found in [3].

Table 1 summarizes the implementation results for all three design approaches, i.e., the top-level specification mapping, the model using a static hierarchical approach, and the model exploiting the proposed partially global optimizations exploiting algorithm 11 and 12. As it can be easily seen, only the design approach exploiting the partially global optimizations of the hCDM leeds to an implementation, which meets the real-time requirements on top of this highly restricted SoC platform.

Table 1. Comparison of the Implementation Results for the FPSLIC SoC

Optimization	**AVR Clock Cycles**	**FPGA CLB**	**FPGA Clock**	**Time per Frame**
High-Level	702500	>100%	-	1s (AVR)
Local	350619	80%	4MHz	74ms
Global	350619	60%	12MHz	21ms

In addition to this real-life example generic specification models were investigated to demonstrate the composition of tasks into models on higher levels of abstraction and the resulting quality. Therfore, two sequential tasks of the randomly generated models are combined into one single task. Again, the genetic design space exploration is taken to optimize the run-time of the resulting implementation. By using ten different specifications consisting of 60 to 100 tasks the average run-times of the reduced models increased by about 10%. But at the same time the number of tasks decreases by a factor of two and so the execution time of the system-level synthesis is reduced in the same manner as depicted in Figure 14.

On the one hand the flexible approach of the hCDM can be used to refine the models, if needed, to reach the design goals. On the other hand parts of the specification can be merged to save design time. It is an important and interesting field of research to find rules and recommendations which party to merge and which parts to refine. Up to now, no general answer can be given.

6 Conclusion

This paper introduces a formal hierarchical specification method, which is aimed to the management of design complexity of embedded systems. Partial models for system-level synthesis can be generated on demand on any level of detail. So, at each point in time only the current necessary complexity of the specification has to be taken into account and, therefore, the whole design process can be flexibly structured in a hierarchical manner. In contrast to specialized hierarchical synthesis algorithms the proposed derivation of flat current model views allows for the application of standard synthesis algorithms. IO Relations are advocated as a very general and flexible means of description for data- and dynamic control-flow behavior on different levels of detail. The proposed method captures the externally visible behavior of IP cores thus hiding internal implementation details. It is therefore very well-suited for IP exploitation. The encapsulation of IP cores by merging IO Relations according to the proposed algorithms reduces considerably both the descriptive complexity and the execution times of subsequent system-level synthesis steps. A subject for future work is to determine cases and reasons, which allow encapsulation of modules, without affecting the resulting accuracy of the generated IP core.

References

1. Atmel, I.: Configurable logic data book (2001), `http://www.atmel.com`
2. Berry, G., Gonthier, G.: The Esterel Synchronous Programming Language: Design, Semantics, Implementation. Science of Computer Programming 19(2), 87–152 (1992)
3. Bieger, J., Huss, S., Jung, M., Klaus, S., Steininger, T.: Rapid prototyping for configurable system-on-a-chip platforms: A simulation based approach. In: IEEE Proc. 17th Int. Conf. on VLSI Design and 3rd Int. Conf. on Embedded Systems. IEEE Computer Society Press, Mumbai (2004)
4. Bossung, W., Huss, S.A., Klaus, S.: High-Level Embedded System Specifications Based on Process Activation Conditions. VLSI Signal Processing, Special Issue on System Design 21(3), 277–291 (1999)
5. Buck, J.T.: Scheduling dynamic dataflow graphs with bounded memory using the token flow model. Tech. Rep. ERL-93-69, UC Berkeley (September 1993)
6. Cortés, L.A., Eles, P., Peng, Z.: A survey on hardware/software codesign representation models. Tech. rep., Linköping University (June 1999)
7. Dave, B., Jha, N.: Cohra: Hardware-software co-synthesis of hierarchical distributed embedded system architectures. In: Proc. Eleventh Int. Conference on VLSI Design. IEEE Computer Society Press, Chennai (1998)

8. Eles, P., Kuchcinski, K., Peng, Z., Doboli, A., Pop, P.: Scheduling of conditional process graphs for the synthesis of embedded systems. In: Proceedings of the conference on Design, automation and test in Europe, pp. 132–139. IEEE Computer Society Press, Paris (1998)
9. Ernst, R., Ziegenbein, D., Richter, K., Thiele, L., Teich, J.: Hardware/Software Codesign of Embedded Systems - The SPI Workbench. In: Proc. IEEE Workshop on VLSI. IEEE Computer Society Press, Orlando (1999)
10. Gajski, D.D., Vahid, F., Narayan, S., Gong, J.: Specification and Design of Embedded Systems. Prentice Hall, Upper Saddle River (1994)
11. Harel, D.: Statecharts: A visual formalism for complex systems. Science of Computer Programming 8(3), 231–274 (1987)
12. Haubelt, C., Mostaghim, S., Slomka, F., Teich, J., Tyagi, A.: Hierarchical Synthesis of Embedded Systems Using Evolutionary Algorithms. In: Evolutionary Algorithms for Embedded System Design. Genetic Algorithms and Evolutionary Computation (GENA), pp. 63–104. Kluwer Academic Publishers, Boston (2003)
13. Jakoby, A., Liskiewicz, M., Reischuk, R.: The expressive power and complexity of dynamic process graphs. In: Workshop on Graph-Theoretic Concepts in Computer Science. pp. 230–242 (2000)
14. Jhumka, A., Klaus, S., Huss, S.A.: A dependability-driven system-level design approach for embedded systems. In: Proceedings of IEEE/ACM International Conference on Design Automation and Test in Europe (DATE), Munich, Germany (March 2005)
15. Klaus, S., Laue, R., Huss, S.A.: Design space exploration of incompletely specified embedded systems by genetic algorithms. In: GI/ITG/GMM Workshop Modellierung und Verifikation, München (April 2005)
16. Klaus, S., Huss, S.A.: Interrelation of Specification Method and Scheduling Results in Embedded System Design. In: Proc. ECSI Int. Forum on Design Languages, Lyon, France (September 2001)
17. Kountouris, A.A., Wolinski, C.: Efficient scheduling of conditional behaviours for high-level synthesis. ACM Transactions on Design Automation of Electronic Systems 7(3), 380–412 (2002)
18. Lee, E.A., Messerschmitt, D.G.: Synchronous dataflow. Proceedings of the IEEE 75(9), 1235–1245 (1997)
19. Mohanty, S., Prasanna, V.K., Neema, S., Davis, J.: Rapid design space exploration of heterogeneous embedded systems using symbolic search and multi-granular simulation. In: Proceedings of the Joint Conference on Languages, Compilers and Tools for Embedded Systems, pp. 18–27. ACM Press, Berlin (2002)
20. Petri, C.A.: Interpretations of a net theory. Tech. Rep. 75–07, GMD, Bonn, W. Germany (1975)
21. Ziegenbein, D., Richter, K., Ernst, R., Thiele, L., Teich, J.: SPI- a system model for heterogeneously specified embedded systems. IEEE Trans. on VLSI Systems 9(4), 379–389 (2002)

Side-Channel Analysis – Mathematics Has Met Engineering

Werner Schindler[1,2]

[1] Bundesamt für Sicherheit in der Informationstechnik (BSI)
Godesberger Allee 185–189
53175 Bonn, Germany
Werner.Schindler@bsi.bund.de
[2] CASED (Center for Advanced Security Research Darmstadt)
Mornewegstraße 32
64289 Darmstadt, Germany
Werner.Schindler@cased.de

Dedicated to Sorin Huss on the occasion of his 60th birthday

Abstract. We illustrate the relevance of advanced mathematical methods in side-channel analysis by two detailed examples. This emphasizes the central statement of this paper that progress in the field of side-channel analysis demands a close cooperation between mathematicians and engineers.

Keywords: Side-channel analysis, timing attack, power attack, secure design, mathematics, engineering science, stochastic process, multivariate statistics.

1 Introduction

In 1996 Paul Kocher introduced timing attacks [16] and in 1999 power analysis [17]. Both papers virtually electrified the community because until then it had appeared as if security analysis of cryptographic devices could be broken up into several independent processes, which can be treated separately by mathematicians (cryptographic algorithms and protocols), computer scientists (operating systems, communication protocols) and engineers (hardware security). In the first years side-channel analysis essentially was a research domain for engineers. Especially in the case of power analysis this seemed to be natural because electrical engineers have expertise in hardware but not mathematicians or computer scientists, and they were able to perform experiments.

As a natural consequence, the applied mathematical methods usually were elementary, and often only a fraction of the available information was exploited. Moreover, when a countermeasure had been found that prevented (or, at least, seemed to prevent) an attack the community usually lost its interest in this attack and in deeper analysis. It is, however, absolutely desirable to know the strongest

A. Biedermann and H. Gregor Molter (Eds.): Secure Embedded Systems, LNEE 78, pp. 43–62.
springerlink.com

attacks against cryptosystems, not only for attackers but also for designers and evaluators. Otherwise it is not clear whether the implemented countermeasures are really strong and reliable, i.e. how to assess countermeasures in a trustworthy way. In particular, one should always understand the 'source' of an attack and not only learn how to execute it.

In the meanwhile the situation has definitely turned to the better, and more sophisticated mathematical methods have been introduced, although the cooperation between mathematicians and security engineers might still be improved. At present some researchers even try to tackle side-channel attacks with formal methods or by the notion of information theory. In this regard side-channel analysis is in line with the physics, mechanics, astronomy and engineering sciences, which have had close and fruitful cooperations with mathematics for centuries, influencing each other heavily. Many branches of mathematics would presumably not have been developed or at least would not have their relevance without these appplications, consider, e.g., analysis or partial differential equations.

In this paper we discuss intensively two examples that show the fruitful connection between mathematics and engineering science in side channel analysis.

The first example treats a well-known timing attack [9], whose efficiency was improved by factor 50 by the formulation and analysis of stochastic processes. Maybe more important, the gained insights smoothed the way for several new attacks in the following years.

The second example is maybe even more instructive. It considers the stochastic approach in power analysis, a well-known effective and efficient attack method that combines engineering's expertise with advanced stochastic methods from multivariate statistics. Moreover, the stochastic approach quantifies the leakage with respect to a vector space basis, a feature that can constructively be used for secure design.

2 My Personal Relation to Side-Channel Analysis

The way I came into contact with side-channel analysis is rather unusual. In 1998 I received my postdoctoral qualification (Privatdozent) and in the summer term 1999 I gave my first course at the mathematical department of Darmstadt University of Technology. In my habilitation thesis [24] (extended to the monograph [30] later) I had considered problems from the field of measure and integration theory with impact on analysis, conclusive statistics and the generation of non-uniformly distributed pseudorandom numbers. So far, my scientific life had mainly been coined by stochastics and measure theory and by several fields from pure mathematics. On the other hand, I had already worked at BSI for almost six years, and so I decided to give a course entitled "Selected topics in modern cryptography". Finally, I had a 'gap' of one and a half 90-minute lectures to fill. I remembered a timing attack on an early version of the Cascade chip, which had been presented at CARDIS 1998 one year before [9]. So I decided to treat this paper in my course.

To make it short, after having studied [9] I was convinced that it is possible to improve this attack significantly. This was the way I came to side-channel analysis, and I have been stuck to this field till today. The preparation of a lesson was the keystone for many research cooperations with engineers working in the field of IT security, also with Sorin Huss. In 2010 we jointly introduced the new international workshop COSADE (Constructive Side-Channel Analysis and Secure Design), which is intended to take place annually in the future.

It should be mentioned that the jubilar Sorin Huss appreciates the use of efficient mathematical methods in engineering science and also has own experience. In his PhD thesis [12] he applied mathematical methods to develop integrated circuits. His development method aims at the optimal solution with respect to the trade-off between development costs and efficiency of the design solution, and he demonstrated the applicability of his approach.

3 Timing Attack from CARDIS 1998

Target of the attack was an early version of the Cascade (square & multiply exponentiation algorithm, 'direct' exponentiation (i.e. no Chinese Remainder Theorem (CRT)), Montgomery's multiplication algorithm, ARM7M Risc processor), which was not protected against timing attacks.

The square & multiply exponentiation algorithm computes modular exponentiations by sequences of hundreds or even more than thousand modular squarings and multiplications. As any timing attack exploits input-dependent timing differences of modular multiplications and squarings it is necessary to explicitly consider the modular multiplication algorithm. Montgomery's multiplication algorithm [20], Sect. 14, is widely used since it is very fast. The main reason is that modular reductions can be moved to power-of-2 moduli.

As usually, $Z_M := \{0, 1, \ldots, M-1\}$, and for an integer $b \in Z$ the term $b(\bmod M)$ denotes the unique element of Z_M that is congruent to b modulo M. In the following we assume that M is an odd modulus (e.g., an RSA modulus or a prime factor) and that $R := 2^x > M$ is a power of two (e.g. $x = 512$ or $x = 1024$). For input a, b Montgomery's multiplication algorithm returns $\mathrm{MM}(a, b; M) := abR^{-1}(\bmod M)$.

Usually, a time-efficient multiprecision variant of Montgomery's algorithm is implemented, which is tailored to the device's hardware architecture. Assume that ws denotes the word size for the arithmetic operations (e.g. $ws = 32$) and that ws divides the exponent x. Then $r := 2^{ws}$ and $R = r^v$ with $v = x/ws$ (Example: $x = 512$, $ws = 32$, $v = 16$). For the moment let further $a = (a_{v-1}, \ldots, a_0)_r$, $b = (b_{v-1}, \ldots, b_0)_r$, and $s = (s_{v-1}, \ldots, s_0)_r$ denote the r-adic representations of a, b and s, respectively. The term $R^{-1} \in Z_M$ denotes the multiplicative inverse of R in Z_M, i.e. $RR^{-1} \equiv 1 \pmod M$. The integer $M^* \in Z_R$ satisfies the integer equation $RR^{-1} - MM^* = 1$, and $m^* := M^*(\bmod r)$.

Algorithm 1. Montgomery's algorithm

```
0.) Input: a,b
1.) s:=0
2.) for i=0 to v-1 do {
        u_i := (s_0 + a_i * b_0)m * (mod r)
        s := (s + a_i b + u_i m)/r
    }
3.) if s ≥ M then s := s−M          \* extra reduction step *\
4.) return s   ( = MM(a,b;M) = abR^{-1}(mod M) )
```

At least for smart cards it is reasonable to assume that for fixed parameters M, R and r the execution time for Step 2 are identical for all pairs of operands (a, b). Timing differences of single modular multiplications or squarings are thus only caused by the fact whether in Step 3 an integer substraction, a so-called *extra reduction* (abbreviated by *ER* in the following), has to be carried out or not. Consequently, in the following we may assume

$$\text{Time}\,(\text{MM}(a, b; M)) \in \{c, c + c_{\text{ER}}\} \tag{1}$$

Remark 1. Security software on PCs as OpenSSL, for instance, may process small operands (i.e., those with leading zero-words) faster due to optimizations of the underlying integer multiplication algorithm. This effect would be irrelevant for the type of timing attack that is considered in this section anyway but may cause additional difficulties for certain chosen-input attacks, cf. [5,1].

Algorithm 2 combines the square & multiply algorithm (or, shortly: s&m algorithm) with Montgomery's multiplication algorithm at cost of two additional modular multiplications (compared to other modular multiplication algorithms).

Algorithm 2. s&m exponentiation with Montgomery's multiplication algorithm

```
0.) Input: y
    (returns y^d(mod n) for d = (d_{w-1} = 1, d_{w-2}, ..., d_0)_2)
1.) ȳ = MM(y, R^2; n)  (= yR(mod n))    (pre-operation)
2.) temp := ȳ
    for k=w-1 downto 0 do {
      temp := MM(temp, temp; n)
      if (d_k = 1) then temp := MM(temp, ȳ; n)
    }
3.) return temp := MM(temp, 1; n)  (= y^d(mod n); post-operation)
```

3.1 The Original Attack [9]

At first the attacker measures the execution times $t_1 := \text{Time}(y_{(1)}^d(\text{mod}\, n), \ldots,$ $t_N := \text{Time}(y_{(N)}^d(\text{mod}\, n)$. More precisely, he gets $\widetilde{t}_1 := t_1 + t_{err(1)}, \ldots, \widetilde{t}_N :=$ $t_N + t_{err(N)}$ where $t_{err(j)}$ denotes the measurement error for basis $y_{(j)}$ (which hopefully is small compared to the effects we want to exploit). The attacker

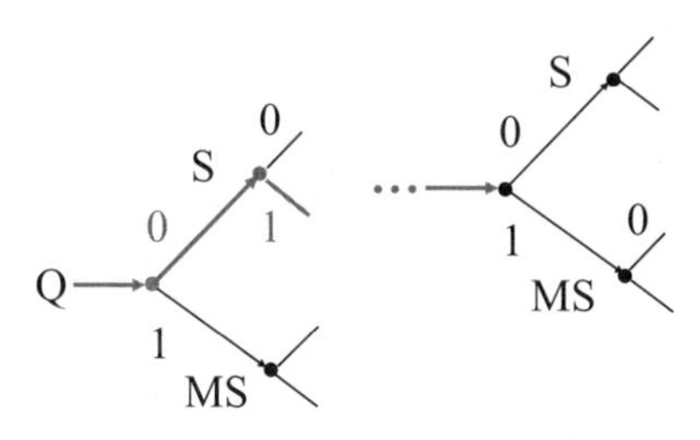

Fig. 1. The correct path in the tree corresponds to the correct key

guesses the secret exponent bits $d_{w-1} = 1, d_{w-2}, \ldots, d_0$ bit by bit, beginning with d_{w-2}. Now assume that he has already guessed the exponent bits $d_{w-1} = 1, d_{w-2}, \ldots, d_{k+1}$ correctly. The next task is to guess exponent bit d_k. The whole attack corresponds to finding a path in a tree. Fig. 1 illustrates the situation. The letters 'M' and 'S' stand for a multiplication of the temp variable in Algorithm 2 with the transformed basis $\bar{y}$ and for a squaring of temp, respectively.

If the attacker has guessed the exponent bits $d_{w-1}, \ldots, d_{k+1}$ correctly (our assumption from above) he can calculate the intermediate temp-values $\text{temp}_1, \ldots, \text{temp}_N$ for all bases $y_{(1)}, \ldots, y_{(N)}$ at this stage. Reference [9] sets

$$\begin{aligned} \mathcal{M}_1 &:= \{y_{(j)} \mid \text{ER in MM}(\text{temp}_j, \bar{y}_{(j)}; n)\} \qquad (2)\\ \mathcal{M}_2 &:= \{y_{(j)} \mid \text{no ER in MM}(\text{temp}_j, \bar{y}_{(j)}; n)\}\\ \mathcal{M}_3 &:= \{y_{(j)} \mid \text{ER in MM}(\text{temp}_j, \text{temp}_j; n)\}\\ \mathcal{M}_4 &:= \{y_{(j)} \mid \text{no ER in MM}(\text{temp}_j, \text{temp}_j; n)\} \quad \text{and}\\ \widetilde{\mu}_i &:= \sum_{y_{(j)} \in \mathcal{M}_i} \widetilde{t}_j / |\mathcal{M}_i| \quad \text{(average execution time) for } i = 1, \ldots, 4. \qquad (3) \end{aligned}$$

We first note that $\mathcal{M}_1 \cup \mathcal{M}_2 = \mathcal{M}_3 \cup \mathcal{M}_4 = \{y_{(1)}, \ldots, y_{(N)}\}$. If $(d_k = 1)$ then $\text{MM}(\text{temp}_j, \bar{y}_{(j)}, n)$ is the next operation in the computation of $y_{(j)}^d (\bmod\ n)$ while $\text{MM}(\text{temp}_j, \text{temp}_j, n)$ is not part of the exponentiation. Thus we may expect that $\widetilde{\mu}_1 > \widetilde{\mu}_2$ and $\widetilde{\mu}_3 \approx \widetilde{\mu}_4$ if $(d_k = 1)$. Of course, whether $\text{MM}(\text{temp}_j, \text{temp}_j; n)$ requires an extra reduction is irrelevant for the exponentiation time t_j. For $d_k = 0$ the roles of $(\widetilde{\mu}_1, \widetilde{\mu}_2)$ and $(\widetilde{\mu}_3, \widetilde{\mu}_4)$ swap. This leads to the diffuse guessing strategy from [9]

$$\begin{aligned} \widetilde{d}_k &:= 1 \quad \text{if } (\widetilde{\mu}_1 \underset{\text{(clearly)}}{>} \widetilde{\mu}_2) \text{ and } (\widetilde{\mu}_3 \approx \widetilde{\mu}_4) \qquad (4)\\ \widetilde{d}_k &:= 0 \quad \text{if } (\widetilde{\mu}_1 \approx \widetilde{\mu}_2) \text{ and } (\widetilde{\mu}_3 \underset{\text{(clearly)}}{>} \widetilde{\mu}_4) \end{aligned}$$

where $\widetilde{d}_k$ denotes the guess for the exponent bit d_k.

Practical Experiments 1. (cf. [9,27]) The timing measurements were performed with an emulator that predicts the execution time of a program in clock cycles. The code was the ready-for-transfer version of the Cascade library, i.e. with critical routines directly written in the card's native assemble language. Since the emulator is designed to allow implementors to optimise their code before 'burning' the actual smart cards, its predictions should match almost perfectly the properties of the 'real' implementation. The attack from [9] required 200.000 – 300.000 measurements to recover a 512-bit key.

The experiments verify that the guessing strategy (4) definitely works. On the negative side the decision strategy exploits only a fraction of the information. The equivalent path-finding problem depicted in Fig. 1 illustrates that there is no chance to return to the right path after any wrong decision. A successful attack requires $\lfloor \log_2(n) \rfloor$ correct decisions. In other words: The decision strategy should exploit all the information that is available.

As already mentioned I was sure that the original attack [9] was far from being optimal. The first task was to analyse Montgomery's multiplication algorithm.

3.2 A Closer Look on Montgomery's Multiplication Algorithm

The central task is to understand the timing behaviour of Montgomery's multiplication algorithm within modular exponentiation (Algorithm 2). We summarise results from [25,28,31] on the stochastic behaviour of the extra reductions.

Definition 1. *Random variables are denoted with capital letters, their realisations (values assumed by these random variables) with the respective small letters. The term $N(\mu, \sigma^2)$ denotes a normal distribution with mean μ and variance σ^2.*

Lemma 1(i) says that whether $\mathrm{MM}(a, b; M)$ requires an ER depends only on a, b, M, R but not on the word size ws. This is very helpful as it allows to assume $x = 1$ in the proof Assertion (ii). Of course, word size ws and the concrete architecture influence the values c, c_{ER} but this is irrelevant for the moment.

Lemma 1. *(i) For each word size ws for Algorithm 1 the intermediate result after Step 2 in Algorithm 2 equals $s = (ab + uM)/R$ with $u = abM^* (\mathrm{mod}\, R)$.*
(ii) $\frac{\mathrm{MM}(a,b;M)}{M} = \left(\frac{a}{M} \frac{b}{M} \frac{M}{R} + \frac{abM^* \quad (\mathrm{mod}\, R)}{R} \right) (\mathrm{mod}\, 1)$. That is, an extra reduction is carried out iff the sum within the bracket is ≥ 1 iff $\frac{\mathrm{MM}(a,b;M)}{M} < \frac{a}{M} \frac{b}{M} \frac{M}{R}$.

Proof. [25], Remark 1, [31], Lemma 2.

Exponentiation Algorithm 2 initialises the variable temp with the pre-processed base $\bar{y}$ and computes a sequence of modular squarings until $\text{temp} = y^d (\mathrm{mod}\, n)$ is computed. We interpret the *normalised intermediate values* from the exponentiation of basis $y_{(j)}$, given by $\text{temp}_{0(j)}/n, \text{temp}_{1(j)}/n, \ldots \in [0, 1)$, as realizations of $[0, 1)$-valued random variables $S_{0(j)}, S_{1(j)}, \ldots$. Consequently, we interpret the time for the ith Montgomery operation (squaring of $\text{temp}_{i(j)}$ or multiplication of $\text{temp}_{i(j)}$ by $\bar{y}$) as a realization of $c + W_{i(j)} \cdot c_{\text{ER}}$, where $W_{i(j)}$ is a $\{0,1\}$-valued random variable for all i, j, assuming 1 iff an extra reduction is necessary. Understanding the stochastic process $W_{1(j)}, W_{2(j)}, \ldots$ is necessary to determine the optimal decision strategy.

From Lemma 1 we deduce the following relations where the right-hand sides differ for the two possible types of Montgomery operations in Step i. Again '(j)' refers to the base $y_{(j)}$.

$$S_{i+1(j)} := \begin{cases} \frac{n}{R} S_{i(j)}^2 + V_{i+1(j)} (\text{mod } 1) & \text{for MM(temp, temp}; n) \\ \frac{\bar{y}}{n} \frac{n}{R} S_{i(j)} + V_{i+1(j)} (\text{mod } 1) & \text{for MM(temp}, \bar{y}; n) \end{cases} \tag{5}$$

Lemma 2(ii) in [31] says that we may assume that the random variables $V_{1(j)}, V_{2(j)}, \ldots$ are iid equidistributed on $[0,1)$. As an immediate consequence, the random variables $S_{1(j)}, S_{2(j)}, \ldots$ are also iid equidistributed on $[0,1)$, which is hardly surprising. The point here is that the random variables $W_{1(j)}, W_{2(j)}, \ldots$, which describe the (random) timing behaviour of the Montgomery operations within the exponentiation of $y_{(j)}$ with exponentiation Algorithm 2, can be expressed in terms of the random variables $S_{0(j)}, S_{1(j)}, \ldots$. More precisely,

$$W_{i(j)} := \begin{cases} 1_{S_{i(j)} < S_{i-1(j)}^2 (n/R)} & \text{for MM(temp, temp}; n) \\ 1_{S_{i(j)} < S_{i-1(j)} (\bar{y}_{(j)}/n)(n/R)} & \text{for MM(temp}, \bar{y}_{(j)}; n). \end{cases} \tag{6}$$

The sequence $W_{1(j)}, W_{2(j)}, \ldots$ is neither independent nor identically distributed but $W_{i(j)}$ and $W_{i+1(j)}$ are negatively correlated. On the other hand, the tuples $(W_{i(j)}, W_{i+1(j)}, \ldots, W_{i+m(j)})$ and $(W_{k(j)}, W_{k+1(j)}, \ldots, W_{k+t(j)})$ (but not their components!) are independent if $k > i + m + 1$. In particular, (6) yields the one-dimensional distribution

$$E(W_i(j)) = \begin{cases} \frac{1}{3} \frac{M}{R} & \text{for MM(temp, temp}; n) \\ \frac{1}{2} \frac{\bar{y}_{(j)}}{M} \frac{M}{R} & \text{for MM(temp}, \bar{y}_{(j)}; n). \end{cases} \tag{7}$$

As easily can be seen, the probabilities for an ER in a squaring and in a multiplication with $\bar{y}$ differ. For multiplications this probability depends linearly on $\bar{y}_{(j)}$.

Remark 2. (i) It should be mentioned that this approach can be extended from the square & multiply algorithm to table-based exponentiation algorithms as well (cf. to [20], Alg. 14.79 (fixed windows exponentiation), [20], Alg. 14.85 (sliding windows exponentiation)), which is relevant for the attacks addressed in Subsection 3.4). For details the interested reader is referred to [31], Sect. 3.
(ii) It seems reasonable to follow a similar strategy for a security analysis of other modular multiplication algorithms.

3.3 The Optimised CARDIS Timing Attack

In this subsection we sketch the central steps from [28,27]. As in the original attack the attacker measures the exponentiation times $\widetilde{t}_{(j)} := \text{Time}(y_{(j)}^d \pmod n) + t_{\text{Err(j)}}$ for a sample $y_{(1)}, \ldots, y_{(N)}$. More precisely,

$$\widetilde{t}_{(j)} = t_{\text{Err}(j)} + t_{S(j)} + (v + \text{ham}(d) - 2)c + \left(w_{1(j)} + \ldots + w_{v+\text{ham}(d)-2(j)}\right) c_{\text{ER}} \tag{8}$$

where $\text{ham}(d)$ denotes the Hamming weight of d, and $w_{i(j)} \in \{0,1\}$ equals 1 iff the ith Montgomery operation requires an extra reduction for basis $y_{(j)}$ and 0 else. The term $t_{S(j)}$ summarises the time needed for all operations apart from the Montgomery multiplications in Step 2 (input, output, handling the loop variable, evaluating the if-statements, pre- and post-multiplication). We may assume that the attacker knows $t_{S(j)}$ exactly; possible deviations may be interpreted as part of the measurement error $t_{\text{Err}(j)}$. We further assume that the attacker had already guessed the parameters v, $\text{ham}(d)$, c and c_{ER} in a pre-step of the attack (cf. [26], Sect. 6).

As in Subsection 3.1 we assume that the most significant exponent bits $d_{v-1}, \ldots, d_{k+1}$ have been guessed correctly. We focus on exponent bit d_k. For each basis $y_{(j)}$ the attacker subtracts the execution time that is needed to process the exponent bits $d_{v-1}, \ldots, d_{k+1}$ and $t_{S(j)}$ from the measured overall exponentiation time $\widetilde{t}_j$, which gives the time that is required for the remaining bits $d_k, \ldots, d_0$ (beginning at the step 'if $(d_k = 1)$ then $\text{MM}(\text{temp}_{(j)}, \bar{y}_{(j)}; n)$' in Algorithm 2). From $\text{ham}(d)$ he computes the number m of remaining Montgomery multiplications for $\bar{y}_{(j)}$.

Since time differences for the remaining operations are caused by the number of extra reductions (and maybe to some extent by measurement errors) we consider the 'normalised' remaining execution time

$$\widetilde{t}_{d,rem(j)} := \frac{\widetilde{t}_{(j)} - t_{S(j)} - (v + \text{ham}(d) - 2)c}{c_{\text{ER}}} - \sum_{i=1}^{v+\text{ham}(d)-2-k-m} w_{i(j)} \tag{9}$$

$$= t_{dErr(j)} + \sum_{i=v+\text{ham}(d)-k-m-1}^{v+\text{ham}(d)-2} w_{i(j)}.$$

The right-hand summand in the second line is the number of extra reductions for the remaining Mongomery multiplications. The remaining Montgomery operations are labelled by indices $v + \text{ham}(d) - k - m - 1, \ldots, v + \text{ham}(d) - 2$. The normalised measurement error $t_{dErr(j)} = t_{Err(j)} / c_{\text{ER}}$ is interpreted as a realization of an $N(0, \alpha^2 = \sigma_{Err}^2 / c_{\text{ER}}{}^2)$-distributed random variable that is independent of the sequence $W_{1(j)}, W_{2(j)} \ldots$ (cf. [28], Sect. 6).

To determine the optimal attack strategy we interpret the guessing of exponent bit d_k as a statistical decision problem $(\Theta = \{0,1\}, \Omega, s\colon \Theta \times A \to [0,\infty), \text{DE}, A = \{0,1\})$. The set Θ contains the admissible parameters (here: possible values of d_k, i.e. 0 and 1), and the attacker decides for some value in A (admissible decisions) on basis of an observation $\omega \in \Omega$. The loss function $s(\theta, a)$ quantifies the loss if $\theta \in \Theta$ is correct but the attacker decides for $a \in A$. The attacker may apply any τ from $\text{DE} := \{\tau'\colon \Omega \to A\}$, the set of all (deterministic) decision strategies. We refer the reader to [31], Sect. 2, which concentrates on finite sets $\Theta = A$, the case that is relevant for side-channel analysis. For a more general treatment of statistical decision theory cf. [18], for instance.

A false guess $\widetilde{d}_k \neq d_k$ implies wrong assumptions on the intermediate temp values for both hypotheses $d_t = 0$ and $d_t = 1$ for all the forthcoming decisions (when guessing d_t for $t < k$). Consequently, these guesses cannot be reliable, and hence we use the loss function $s(0,1) = s(1,0) = 1$, and $s(0,0) = s(1,1) = 0$ since a correct decision should not imply a loss. Further useful information is the probability distribution of the exponent bit d_k (the so-called a priori distribution). If the secret exponent d has been selected randomly or if the public exponent e is fixed it is reasonable to assume that $\eta(1) := \text{Prob}(d_k = 1) = (m-1)/k$ since $d_0 = 1$. That is, the a priori distribution for bit d_k is given by $(\eta(0), \eta(1)) = ((k+1-m)/k, (m-1)/k)$.

If $d_k = 1$ the next Montgomery operations in the computation of $y_{(j)}^d (\text{mod } n)$ are $\text{MM}(\text{temp}_j, \bar{y}_{(j)}; n)$ and $\text{MM}(\text{MM}(\text{temp}_j, \bar{y}_{(j)}; n), \text{MM}(\text{temp}_j, \bar{y}_{(j)}; n); n)$ (multiplication and squaring) while if $d_k = 0$ the next Montgomery operation is $\text{MM}(\text{temp}_j, \text{temp}_j; n)$ (squaring). The variables $u_{M(j)}, u_{S(j)}, t_{S(j)} \in \{0,1\}$ are 1 iff the respective operation requires an ER. That is, $u_{M(j)}$ and $u_{S(j)}$ are summands of the right-hand side of (9) if $\theta = 1$ whereas $t_{S(j)}$ is such a summand if $\theta = 0$. The attacker decides on basis of the 'observation' $\left(\tilde{t}_{\text{drem}(j)}, u_{M(j)}, u_{S(j)}, t_{S(j)}\right)_{j \leq N}$. The stochastic process $W_{1(j)}, W_{2(j)}, \ldots$ quantifies the (random) timing behaviour of the Montgomery multiplications that are determined by $d_{w-2}, \ldots, d_0$. These random variables are neither independent nor stationary distributed. However, under weak assumptions they meet the assumptions of a version of the central limit theorem for dependent random variables ([28], Lemma 6.3(iii)). Since $W_{i(j)}$ and $W_{r(j)}$ are independent if $|i-r| > 1$ (cf. Subsect. 3.2) we further conclude

$$\text{Var}\left(W_{1(j)} + \ldots + W_{t(j)}\right) = \sum_{i=1}^{t} \text{Var}(W_{i(j)}) + 2\sum_{i=1}^{t-1} \text{Cov}(W_{i(j)}, W_{i+1(j)}) \qquad (10)$$

For the variances we have to distinguish between two cases (squaring, multiplication with $\bar{y}_{(j)}$; cf. (6)), for the covariances between three cases, where $W_{i(j)}$ and $W_{i+1(j)}$ correspond to two squarings (cov_{SS}), resp. to a squaring followed by a multiplication with $\bar{y}_{(j)}$ ($\text{cov}_{\text{SM(j)}}$), resp. to a multiplication with $\bar{y}_{(j)}$ followed by a squaring ($\text{cov}_{\text{MS(j)}}$). By (6) the random vector $(W_{i(j)}, W_{i+1(j)})$ can be expressed as a function of the iid random variables $S_{i-1(j)}, S_{i(j)}, S_{i+1(j)}$. For instance, $\text{Cov}_{\text{MS}}(W_i W_{i+1}) =$

$$\int_{[0,1)^3} 1_{\{s_i < s_{i-1}\bar{y}_{(j)}/R\}} \cdot 1_{\{s_{i+1} < s_i^2 n/R\}}(s_{i-1}, s_i, s_{i+1})\, ds_{i-1} ds_i ds_{i+1} - \frac{\bar{y}_{(j)}}{2R} \cdot \frac{n}{3R} \qquad (11)$$

Elementary but careful computations yield

$$\text{cov}_{\text{MS(j)}} = 2p_j^3 p_* - p_j p_*, \quad \text{cov}_{\text{SM(j)}} = \frac{9}{5} p_j p_*^2 - p_j p_* \qquad (12)$$
$$\text{cov}_{\text{SS}} = \frac{27}{7} p_*^4 - p_*^2 \qquad \text{with } p_j := \frac{\bar{y}_{(j)}}{2R} \text{ and } p_* := \frac{n}{3R}.$$

Since the random variables $W_{1(j)}, W_{2(j)}, \ldots$ are not independent the distribution of $W_{i+1(j)} + \cdots + W_{t(j)}$ depends on the preceding value $w_{i(j)}$. Theorem 1 considers this fact. We first introduce some abbreviations.

Notation. $hn(0,j) := (k-1)p_*(1-p_*) + mp_j(1-p_j) + 2(m-1)\mathrm{cov}_{\mathrm{MS(j)}} + 2(m-1)\mathrm{cov}_{\mathrm{SM(j)}} + 2(k-m-1)\mathrm{cov}_{\mathrm{SS}} + 2\frac{k-m}{k-1}\mathrm{cov}_{\mathrm{SM(j)}} + 2\frac{m-1}{k-1}\mathrm{cov}_{\mathrm{SS}} + \alpha^2$,
$hn(1,j) := (k-1)p_*(1-p_*) + (m-1)p_j(1-p_j) + 2(m-2)\mathrm{cov}_{\mathrm{MS(j)}} + 2(m-2)\mathrm{cov}_{\mathrm{SM(j)}} + 2(k-m)\mathrm{cov}_{\mathrm{SS}} + 2\frac{k-m+1}{k-1}\mathrm{cov}_{\mathrm{SM(j)}} + 2\frac{m-2}{k-1}\mathrm{cov}_{\mathrm{SS}} + \alpha^2$,
$ew(0,j \mid b) := (k-1)p_* + mp_j + \frac{k-m}{k-1}(p_{*S(b)} - p_*) + \frac{m-1}{k-1}(p_{jS(b)} - p_j)$,
$ew(1,j \mid b) := (k-1)p_* + (m-1)p_j + \frac{k-m+1}{k-1}(p_{*S(b)} - p_*) + \frac{m-2}{k-1}(p_{jS(b)} - p_j)$ with $p_{*S(1)} := \frac{27}{7}p_*^3$, $p_{*S(0)} := \frac{p_* - p_* p_{*S(1)}}{1-p_*}$, $p_{jS(1)} := \frac{9}{5}p_* p_j$ and $p_{jS(0)} := \frac{p_j - p_* p_{jS(1)}}{1-p_*}$.

For a complete proof of Theorem 1 we refer the interested reader to [28], Theorem 6.5 (i). We mention that the a priori distribution and the definition of the loss function has impact on the optimal decision strategy.

Theorem 1. *(Optimal decision strategy) Assume that the guesses* $\widetilde{d}_{v-1}, \ldots, \widetilde{d}_{k+1}$ *are correct and that* $\mathrm{ham}(d_k, \ldots, d_0) = m$. *Let*

$$\psi_{N,d} : (\mathbb{R} \times \{0,1\}^3)^{\mathrm{N}} \to \mathbb{R}, \qquad \psi_{\mathrm{N,d}}((\widetilde{t}_{\mathrm{drem(1)}}, u_{\mathrm{M(1)}}, \ldots, u_{\mathrm{S(N)}}, t_{\mathrm{S(N)}})) :=$$
$$-\frac{1}{2}\sum_{j=1}^{N}\left(\frac{\left(\widetilde{t}_{\mathrm{drem}(j)} - t_{S(j)} - ew(0,j \mid t_{S(j)})\right)^2}{hn(0,j)} - \frac{\left(\widetilde{t}_{\mathrm{drem}(j)} - u_{M(j)} - u_{S(j)} - ew(1,j \mid u_{S(j)})\right)^2}{hn(1,j)}\right).$$

Then the deterministic decision strategy τ_d: $(\mathbb{R} \times \{0,1\}^3)^{\mathrm{N}} \to \{0,1\}$, *defined by*

$$\tau_d = 1_{\psi_{N,d} < \log\left(\frac{m-1}{k-m+1}\right) + \frac{1}{2}\sum_{j=1}^{N}\log(1+c_j)} \quad \text{with } c_j := \frac{hn(0,j) - hn(1,j)}{hn(1,j)} \tag{13}$$

is optimal (i.e., a Bayes strategy against the a priori distribution η*).*

Sketch of the proof. To apply Theorem 1 (iii) from [31] we first have to determine the conditional probability densities $h_{\theta,*j|C_j}(\widetilde{t}_{\mathrm{drem}(j)}, u_{M(j)}, u_{S(j)}, t_{S(j)})$ (normal distribution) of the random vectors $\boldsymbol{X}_j := (\widetilde{T}_{\mathrm{drem}(j)}, U_{M(j)}, U_{S(j)}, T_{S(j)})$ for $\theta = 0,1$ and $j \le N$ with $C_j = (U_{M(j)} = u_{M(j)}, U_{S(j)} = u_{S(j)}, T_{S(j)} = t_{S(j)})$. (We point out that the $\boldsymbol{X}_j$ are independent but not their components.) The products $\prod_{j=1}^{N} h_{\theta,*j|C_j}(\cdot)$ are inserted in equation (4) from [31], and elementary computations complete the proof of Theorem 1.

The overall attack is successful iff all the guesses $\widetilde{d}_{v-1}, \ldots, \widetilde{d}_0$ are correct. Theorem 6.5 (ii) in [28] quantifies the probability for individual wrong guesses. Guessing errors will presumably only occur in the first phase of the attack since the variance of the sum $W_{v+\mathrm{ham}(d)-k-m-1(j)} + \ldots + W_{v+\mathrm{ham}(d)-2(j)}$ decreases with index k.

Example 1. ([31], Example 3) Assume that the guesses $\widetilde{d}_{v-1}, \ldots, \widetilde{d}_{k+1}$ have been correct. For randomly chosen bases $y_{(1)}, \ldots, y_{(N)}$, for $n/R = 0.7$, $\alpha^2 = 0$, $N \ge 5000$, and ...

(a) ... $(k, m) = (510, 255)$ we have $\text{Prob}(\widetilde{d_k} \neq d_k) \leq 0.014$.
(b) ... $(k, m) = (440, 234)$ we have $\text{Prob}(\widetilde{d_k} \neq d_k) \leq 0.010$.
(c) ... $(k, m) = (256, 127)$ we have $\text{Prob}(\widetilde{d_k} \neq d_k) \leq 0.001$.

Remark 3. The attack falls into a sequence of statistical decision problems (one for each exponent bit). The $\psi_{N,d}$-values themselves may be interpreted as realizations of random variables $Z_{v-1}, Z_{v-2}, \ldots$, which have the pleasant property that their distributions change noticeably after the first wrong guess. For instance, the decision strategy from Theorem 1 then guesses $\widetilde{d_k} = 1$ only with a probability of around 0.20. In Section 3 of [27] a new stochastic strategy was introduced that detects, locates and corrects guessing errors, which additionally reduces the sample size by about 40%.

Practical Experiments 2. (cf. [27]) For the same hardware as for the original attack (cf. Practical Experiments 1) the optimal decision strategy from Theorem 1 (combined with the error detection strategy mentioned above) yielded a success rate of 74% ('real-life' case) for the overall attack for sample size $N = 5000$. In the ideal case, realised by a stochastic simulation, the attacker knows $c, c_{\text{ER}}, v, \text{ham}(d)$ and $t_{(S)}$ exactly. In the ideal case the success rate increased to 85%. For $N = 6000$ we obtained success rates of 85% and 95% (ideal case), respectively. The optimal decision strategy increases the efficiency of the original attack by factor 50. Moreover, the success rates for the 'real-life' case and the ideal case are of the same order of magnitude, another indicator that the stochastic model fits well.

3.4 Stochastic Properties of Montgomery's Multiplication Algorithm: Further Consequences

First of all, the analysis of the stochastic behaviour of Montgomery's multiplication algorithm allowed to improve the original CARDIS attack by factor 50. Both the original attack and its optimization can be prevented by blinding techniques ([16], Sect. 10). However, it is always profitable to understand the source of an attack since countermeasures may only prevent a special type of the attack. The principal weakness clearly is the extra reduction of Montgomery's algorithm. Even more important, Subsection 3.2 collects important properties, which might be used to exploit this feature.

Equation (7) says $\text{Prob}(\text{ER in MM}(\text{temp}_j, \text{temp}_j; n)) = n/3R$ while Prob(ER in $\text{MM}(\text{temp}_j, \bar{y}_{(j)})$) depends linearly on $\bar{y}_{(j)}$. Reference [25] introduces a new timing attack on RSA with CRT and Montgomery's algorithm, which had believed to be immune against timing attacks. The central idea is that an attacker can decide on basis of timing difference $\text{Time}((u_2(R^{-1}(\text{mod}\, n)))^d(\text{mod}\, n)) - \text{Time}((u_1(R^{-1}(\text{mod}\, n)))^d(\text{mod}\, n))$ whether a multiple of one RSA prime factor lies in the interval $\{u_1+1, \ldots, u_2\}$. The timing attack on RSA with CRT is by one order of magnitude more efficient than the timing attack on RSA without CRT. Under optimal conditions (exact timing measurements) about 300 RSA encryptions suffice to factorise the RSA modulus $n = p_1 p_2$. The attack also applies to table-based exponentiation algorithms, though on cost of efficiency. It should be noted that blinding techniques prevent also this type of timing attack. The timing attack on RSA with CRT [25] would definitely not have been detected without a careful analysis of the original CARDIS attack.

Reference [5] adapts this timing attack on RSA with CRT to the sliding windows exponentiation algorithm ([20], 14.85f.), which made a patch of OpenSSL v.0.9.7 necessary. Brumley and Boneh showed that also software implementations on PCs may be vulnerable against timing attacks and that remote timing attacks are principally feasible. Both threats had been underestimated so far. Despite of [25] basis blinding had been disabled by default in OpenSSL until that time. As response to Brumley and Boneh's timing attack the default settings enabled basis blinding by default from OpenSSL v.0.9.7.b. We note that [1] increases the efficiency of this attack by factor 10 and that base blinding prevents this attack, too.

In [2] an instruction cache attack is introduced that again exploits the ER step in Montgomery's multiplication algorithm but is not prevented by basis blinding. A patch of OpenSSL v.0.9.8.e used a result from Walter [38]. More precisely, if the Montgomery constant R is selected by at least factor 4 larger than the modulus the extra reduction step may be skipped. The intermediate temp values will never become larger than twice the modulus. Interestingly, an attack on a similar exponentiation algorithm had already been analysed in [29]. There the information on the ERs was gained by power analysis in place of an instruction cache attack but the mathematical methods are similar. Both attacks [5,2] underline that the cooperation between hardware engineers and software engineers should be improved.

4 A New Method in Power Analysis

Since [17] lots of papers on power analysis have been published. Most of them consider DPA (Differential Power Analysis) or CPA (Correlation Power Analysis) techniques [4,13,21,23,37], and a large number of countermeasures have been proposed ([6,8,34] etc.). Both DPA and CPA apply selection functions with subkey and plaintext (resp. ciphertext) as arguments. Ideally, for the correct subkey the selection function should be strongly correlated to the power consumption at some time instant t whereas the substitution of a wrong subkey into the selection function should reduce the correlation to ≈ 0. DPA typically applies a Hamming weight model or focuses on a particular bit while CPA is more flexible in the choice of the selection function and applies the correlation coefficient. From an abstract standpoint DPA and CPA are substantially very similar. Both are easy to handle but on the negative side they only exploit a fraction of the available information. In particular, it is not clear how to combine power measurements from different time instants effectively.

Template attacks ([7,3] etc.) overcome this problem. Template attacks consider the power consumption at several time instants $t_1 < \cdots < t_m$. The measured power consumption at $\boldsymbol{t} := (t_1, \ldots, t_m))$ is interpreted as a realization of an m-dimensional random vectors $\boldsymbol{I_t}(x, k)$ whose unknown distribution depends on the subkey k, on some part of the plaintext, resp. on some part of the ciphertext, denoted by x. If the attacked device applies masking techniques, the random vector $\boldsymbol{I_t}(x, z, k)$ also depends on a masking value z (random number).

In the profiling phase the adversary uses an identical training device to obtain estimates for the unknown densities (the so-called *templates*) $f_{x,k}(\cdot)$, resp. $f_{x,z,k}$ if masking techniques are applied. In the attack phase the adversary (attacker, designer or evaluator) performs measurements at the target device. It is usually assumed that the unknown densities are (at least approximately) multidimensional normally distributed. Hence profiling reduces to the estimation of mean values and covariance matrices.

In the attack phase the measurement values are substituted into the estimated densities (maximum likelihood principle). This step is equal to the stochastic approach (cf. to (25) and (26)).

In a 'full' template attack the adversary estimates the densities for all pairs (x, k) or even for all triplets (x, z, k) in the case of masking. For given time instants $t_1, \ldots, t_m$ a full template attack should provide maximum attack efficiency among all attacks that focus on the power consumption at $\boldsymbol{t} = (t_1, \ldots, t_m)$, at least if the sample size for the profiling series is sufficiently large. Clear disadvantage is the gigantic workload in the profiling phase, especially if masking techniques are applied, due to the large number of different templates that have to be generated. For strong implementations these densities differ only slightly, demanding a large number of measurements for each template. Under specific conditions the number of templates might be reduced, e.g. if an attacker aims at a chosen-input attack or if he attacks the target device first. These restrictions clearly limit the applicability of an attack, and attacks on masked implementations still demand a very large number of templates anyway. For evaluation purposes it is a priori not clear whether the selected tuples (x, k), resp. selected triples (x, z, k), are typical. Combining template attacks with DPA [22] may relax this problem but the optimality of the attack efficiency may get lost.

The stochastic approach [32,10,19,33,35,15] combines engineers' expertise with quantitative stochastic methods from multivariate statistics. The stochastic approach is clearly stronger than DPA or CPA, and its attack efficiency is comparable to template attacks whereas the number of profiling measurements is by one to several orders of magnitude smaller. Moreover, the leakage is quantified with respect to a vector space basis which supports target-oriented (re-)design.

It should be noted that attempts have been made to introduce formal methods into side-channel analysis or to apply ideas from information theory [36,11]. We do not consider these topics in this paper.

4.1 The Stochastic Approach

Target algorithms of the stochastic approach are block ciphers. The adversary (designer, evaluator, attacker) guesses the key in small portions, called subkeys, on basis of power measurements at time instants $t_1 < \cdots < t_m$. In this subsection we describe the stochastic approach and explain its most relevant features (cf. [32,33] for details).

Definition 2. *The letter* $k \in \{0,1\}^s$ *denotes a subkey,* $x \in \{0,1\}^p$ *a known part of the plaintext or the ciphertext, respectively, and* $z \in M$ *denotes a masking value. The random variables* X *and* Z *assume plaintext or ciphertext parts and masking values, respectively.*

As for template attacks we interpret a measurement value $i_t := i_t(x,z,k)$ at time $t \in \{t_1,\ldots,t_m\}$ as a realization of a random variable $I_t(x,z,k)$ whose (unknown) distribution depends on the triplet (x,z,k) (masking case), resp. of a random variable $I_t(x,k)$ whose (unknown) distribution depends on the tuple (x,k) (non-masking case). Unlike template attacks the stochastic approach does not estimate the corresponding probability densities directly but applies the stochastic model

$$I_t(x,z,k) = h_t(x,z,k) + R_t. \tag{14}$$

The leakage function $h_t(x,z,k)$ quantifies the deterministic part of the power consumption at time t, which depends on x,z and k. The random variable R_t quantifies the 'noise', which is originated by other operations that are carried out at the same time (possibly depending on other subkeys) and measurement errors. The random variable R_t is assumed to be independent of $h_t(x,z,k)$, and we may assume $E(R_t) = 0$. (Otherwise, replace R_t by $R_t - E(R_t)$ and $h_t(x,z,k)$ by $h_t(x,z,k) + E(R_t)$.) The mathematical model for unmasked implementations

$$I_t(x,k) = h_t(x,k) + R_t \quad \text{(unmasked implementation)} \tag{15}$$

follows immediately from (14) by simply cancelling the second argument 'z'. (The same is true for nearly all results in this subsection.) The functions $h_{t_1}(\cdot,\cdot,\cdot),\ldots,$ $h_{t_m}(\cdot,\cdot,\cdot)$ and the distribution of the noise vector $(R_{t_1},\ldots,R_{t_m})$ are unknown. They are estimated on basis of power measurements on an identical training device (profiling). Profiling consists two steps.

Profiling Step 1: Estimation of $h_t(\cdot,\cdot,\cdot)$. It would be natural to estimate the function values $h_t(x,z,k)$ for all admissible triplets (x,z,k) and all time instants $t \in \{t_1,\ldots,t_m\}$ separately. However, this approach demanded gigantic workload, and we would essentially perform a template attack.

Instead, we estimate *functions* but not function values. Let us fix $t \in \{t_1,$ $\ldots,t_m\}$. For each $k \in \{0,1\}^s$ we consider the restricted function $h_{t;k}\colon \{0,1\}^p \times M \times \{k\} \to \mathbb{R}$, $h_{t;k}(x,z,k) := h_t(x,z,k)$. We view $h_{t;k}$ as an element of a $2^p|M|$-dimensional real subspace $\mathcal{F}_k := \{h'\colon \{0,1\}^p \times M \times \{k\} \to \mathbb{R}\}$. Roughly speaking, the stochastic approach aims at characterizing the relevant source of side-channel leakage, trading maximum precision for workload. Consequently, we aim at the best approximator in a suitably selected subspace $\mathcal{F}_{u,t;k}$ of $\mathcal{F}_k$. The subspace $\mathcal{F}_{u,t;k}$ should be selected with regard to the concrete implementation (cf. [32,10,19,33,35,15] and Example 2 below). $\mathcal{F}_{u,t;k}$ usually does not contain the exact function $h_{t;k}$ but clearly should be close to it.

The subspace $\mathcal{F}_{u,t;k}$ is spanned by u linearly independent, known functions $g_{j,t;k}\colon \{0,1\}^p \times M \times \{k\} \rightarrow \mathbb{R}$, $j = 0,\ldots,u-1$. These basis vectors are related to the expected relevant sources of leakage. More precisely,

$$\mathcal{F}_{u,t;k} := \{h'\colon \{0,1\}^p \times M \times \{k\} \rightarrow \mathbb{R} \mid \mathrm{h}' = \sum_{\mathrm{j}=0}^{\mathrm{u}-1} \beta'_{\mathrm{j}} \mathrm{g}_{\mathrm{j,t;k}} \text{ with } \beta'_{\mathrm{j}} \in \mathbb{R}\}. \quad (16)$$

The best approximator $h^*_{t;k}$ is the image of $h_{t;k}(\cdot,\cdot,k)$ under the orthogonal projection with regard to some L^2 scalar product (= usual Euclidean scalar product for uniformly distributed (X, Z), the case of highest practical relevance).

Theorem 2. *Let $k \in \{0,1\}^s$ denote the correct subkey.*
(i) [Minimum property of $h^*_{t;k}$] *Let $h^*_{t;k} \in \mathcal{F}_{u,t;k}$. Then*

$$E_{X,Z}\left(\left(h_t(X,Z,k) - h^*_{t;k}(X,Z,k)\right)^2\right) = \min_{h' \in \mathcal{F}_{u,t;k}} E_{X,Z}\left(\left(h_t(X,Z,k) - h'(X,Z,k)\right)^2\right) \quad (17)$$

if and only if

$$E\left(\left(I_t(X,Z,k) - h^*_{t;k}(X,Z,k)\right)^2\right) = \min_{h' \in \mathcal{F}_{u,t;k}} E\left(\left(I_t(X,Z,k) - h'(X,Z,k)\right)^2\right). \quad (18)$$

(ii) The functions $h_{t_1;k},\ldots,h_{t_m;k}$ can be estimated separately without loss of information.
(iii) [least square estimation of $h^*_{t;k}$] *Let*

$$A := \begin{pmatrix} g_{0,t;k}(x_1,z_1,k) & \ldots & g_{u-1,t;k}(x_1,z_1,k) \\ \vdots & \ddots & \vdots \\ g_{0,t;k}(x_{N_1},z_{N_1},k) & \ldots & g_{u-1,t;k}(x_{N_1},z_{N_1},k) \end{pmatrix}, \quad (19)$$

be a real-valued $(N_1 \times u)$-matrix, and assume that the components of $\boldsymbol{i}_t := (i_t(x_1,z_1,k),\ldots,i_t(x_{N_1},z_{N_1},k))^T \in \mathbb{R}^{\mathrm{N}_1}$ are power measurements at time t. If the $(u \times u)$-matrix A^TA is regular the normal equation $A^TA\boldsymbol{b} = A^T\boldsymbol{i}_t$ has the unique solution

$$\widetilde{\boldsymbol{b}}^* = (A^TA)^{-1}A^T\boldsymbol{i}_t, \qquad \text{with} \quad \widetilde{b}^* := (\widetilde{b}^*_0,\ldots,\widetilde{b}^*_{u-1}). \quad (20)$$

Then $\widetilde{h}^*_{t;k}(\cdot,\cdot,k) = \sum_{j=0}^{u-1} \widetilde{\beta}^*_{j,t} g_{j,t;k}(\cdot,\cdot,k)$ with $\widetilde{\beta}^*_{j,t;k} := \widetilde{b}^*_j$ is the least square estimate of $h^*_{t;k}$. If N_1 tends to infinity $\widetilde{h}^*_{t;k}$ converges to $h^*_{t;k}$ with probability 1.

Proof. [33], Theorem 2.7(i),(iv), and Theorem 3.2.

Theorem 2 is crucial for the stochastic approach. Assertion 2(i) implies that the image $h^*_{t;k}$ of $h_{t;k}$ can be computed without knowledge of $h_{t;k}$ since $h^*_{t;k}$ fulfils a minimum property in the subspace $\mathcal{F}_{u,t;k}$. This allows to move all statistical computations from the high-dimensional vector space $\mathcal{F}_k$ to a low-dimensional subspace $\mathcal{F}_{u,t;k}$, reducing the required number of power traces tremendously since the number of parameters to be estimated decreases from $\dim(\mathcal{F}_k) = 2^p|M|$ to $\dim(\mathcal{F}_{u,t;k}) = u$. For AES implementations, for instance, typically $\dim(\mathcal{F}_k) = 2^8$ or 2^{16} while 2 and 9 are typical values for $\dim(\mathcal{F}_{u,t;k})$. Theorem 2(iii) provides an explicit formula for the computation of $h^*_{t;k}$.

Example 2. ([15], Sect. 3) AES implementation on an FPGA, no masking, last round. We focus on the second key byte $k_{(2)}$. CMOS technology suggests a distance model. A natural choice for the subspace thus is $\mathcal{F}_{9,t;k} = \langle g_{0,t;k}(x,k), g_{1,t;k}(x,k), \ldots, g_{8,t;k}(x,k)\rangle$ with

$$g_{0,t;k_{(2)}}((x_{(6)}, x_{(2)}), k_{(2)}) = 1$$

$$g_{j,t;k_{(2)}}((x_{(6)}, x_{(2)}), k_{(2)}) = (x_{(6)} \oplus S^{-1}(x_{(2)} \oplus k_{(2)}))_j \quad \text{for } j = 1, \ldots, 8 \tag{21}$$

where $x_{(2)}$ and $x_{(6)}$ denote the second and the sixth ciphertext byte, respectively.

Profiling Step 2: Estimation of the Covariance Matrix

Notation. $\boldsymbol{R_t}$ denotes the random vector $(R_{t_1}, \ldots, R_{t_m})$, and, analogously, $\boldsymbol{I_t}(x,z,k)$, $\boldsymbol{i_t}(x_j, z_j, k)$, $\boldsymbol{h^*_{t;k}}(x,z,k)$ etc. denote m-dimensional vectors. Estimates are assigned by the $\sim$ sign.

As for template attacks we assume that all $\boldsymbol{I_t}(x,z,k)$, and thus the random noise vector $\boldsymbol{R_t} = \boldsymbol{I_t}(x,z,k) - \boldsymbol{h_{t;k}}(x,z,k)$ are multidimensional normally distributed. If $\boldsymbol{R_t}$ has regular the covariance matrix $C = (c_{ij})_{1 \le i,j \le m}$ it has density

$$f_C\colon \mathrm{I\!R}^{\mathrm{m}} \to \mathrm{I\!R} \qquad \mathrm{f_C}(\boldsymbol{y}) = \frac{1}{\sqrt{(2\pi)^{\mathrm{m}} \det \mathrm{C}}} \mathrm{e}^{-\frac{1}{2}\boldsymbol{y}^{\mathrm{T}}\mathrm{C}^{-1}\boldsymbol{y}} \tag{22}$$

(cf. to [14], Subsect. 2.5.1, for instance). As C is unknown it has to be estimated from a set of N_2 fresh m-dimensional measurement vectors $\boldsymbol{i_t}(x_1, z_1, k_1), \ldots,$ $\boldsymbol{i_t}(x_{N_2}, z_{N_2}, k_{N_2})$ from the training device. This gives the estimated density $f_{\tilde{C}}(\cdot)$.

Key Extraction (Attack Phase). If k° is the correct (but unknown) subkey the random vector

$$\boldsymbol{I_t}(x,z,k^\circ) \text{ has density } f_{x,z,k^\circ}\colon \mathrm{I\!R}^{\mathrm{m}} \to \mathrm{I\!R},\ \mathrm{f_{x,z,k^\circ}}(\boldsymbol{y}) := \mathrm{f_C}(\boldsymbol{y} - \boldsymbol{h_t}(\mathrm{x,z,k^\circ})). \tag{23}$$

for each $(x,z) \in \{0,1\}^p \times M$. Of course, the masking value z is unknown in the attack phase. Instead, we view z as a realization of a random variable Z, and we apply the probability density

$$\boldsymbol{I_t}(x, Z, k^\circ) \text{ has density } \bar{f}_{x,k^\circ} := \sum_{z' \in M} \mathrm{Prob}(Z = z') \cdot f_{x,z',k^\circ}\,, \tag{24}$$

usually with $\mathrm{Prob}(Z = z') = 2^{-s}$ for all $z \in M$. The adversary performs N_3 measurements at the target device and obtains power vectors $\boldsymbol{i_t}(x_1, z_1, k^\circ), \ldots,$ $\boldsymbol{i_t}(x_{N_3}, z_{N_3}, k^\circ)$ with unknown masking values $z_1, \ldots, z_{N_3}$ and, of course, with unknown subkey k°. Due to (23) and (24) the adversary decides for that subkey candidate $k^* \in \{0,1\}^s$ that maximises

$$\prod_{j=1}^{N_3} \sum_{z'_j \in M} \mathrm{Prob}(Z = z'_j) \cdot f_{\widetilde{C}} \left(\boldsymbol{i_t}(x_j, z_j, k^\circ) - \widetilde{\boldsymbol{h}}^*_{\boldsymbol{t};k}(x_j, z'_j, k) \right). \tag{25}$$

among all $k \in \{0,1\}^s$ (maximum likelihood estimate). For unmasked implementations (25) simplifies to the product

$$\prod_{j=1}^{N_3} f_{\widetilde{C}} \left(\boldsymbol{i_t}(x_j, k^\circ) - \widetilde{\boldsymbol{h}}^*_{\boldsymbol{t};k}(x_j, k) \right) \tag{26}$$

4.2 The Stochastic Approach Supports Design

The stochastic approach quantifies the leakage at time t with regard to the vector space basis $g_{0,t;k}, \ldots, g_{u-1,t;k}$. Namely,

$$\widetilde{h}^*_{t;k} = \sum_{j=0}^{u-1} \widetilde{\beta}_{j,t;k} g_{j,t;k} \tag{27}$$

where the coefficients $\widetilde{\beta}_{j,t;k}$ were estimated according to (20). Large absolute values $|\beta_{j,t;k}|$ for $j > 0$ point to key-dependent weaknesses in the implementation. Equation (27) gives the designer a hint where to find flaws. Example 3 shows that power information can be used in a constructive way to find implementation weaknesses.

Example 3. (cf. [15] for details) Fig. 2 shows $|\beta_{1,t,k_{(1)}}|, \ldots, |\beta_{8,t,k_{(1)}}|$ exemplarily for the subkey $k_{(1)} = 209$ at some time instant t. The numbers on the horizontal axis refer to the particular β-coefficients, e.g. '4' refers to $\beta_{4,t;k_{(1)}}$. Similarly,

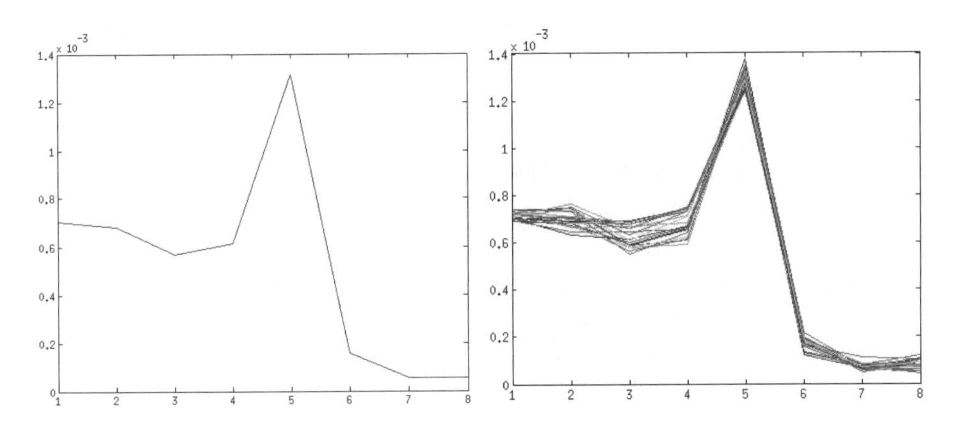

Fig. 2. β-characteristic for subkey $k_{(1)} = 209$ at time t

Fig. 3. β-characteristic for subkey $k_{(1)} = 209$ at time instants $t_1, \ldots, t_{21}$

Fig. 3 plots the β-characteristics for $k_{(1)} = 209$ for time instants $t_1 \ldots, t_{21}$. Differently coloured lines belong to different time instants. The first observation is that the curves are rather similar for all time instants. Further, it is striking that the particular β-coefficients are not almost equal, what seems to be natural. Instead, the absolute values $|\beta_{1,t,k_{(1)}}|, \ldots, |\beta_{4,t,k_{(1)}}|$ are much larger than $|\beta_{6,t,k_{(1)}}|, |\beta_{7,t,k_{(1)}}|, |\beta_{8,t,k_{(1)}}|$, and $|\beta_{5,t,k_{(1)}}|$ is maximal.

In this AES implementation the SBox permutation was realised by a lookup-table, and the design was finally synthesised for the Virtex-II pro family. The β-coefficients indicate that the automatic place & route process treats the particular bits very differently. In fact, a closer look at the design shows that bit 5 switches several first stage multiplexers etc. We refer the interested reader to [15], Sect. 4, for details.

5 Conclusion

Side channel analysis is a fascinating area of research for mathematicians, computer scientists and engineers who are interested in interdisciplinary projects. Especially the second example underlines that further progress is hardly possible without close research cooperations.

References

1. Acıiçmez, O., Schindler, W., Koç, Ç.K.: Improving Brumley and Boneh Timing Attack on Unprotected SSL Implementations. In: Meadows, C., Syverson, P. (eds.) 12th ACM Conference on Computer and Communications Security — CCS 2005, pp. 139–146. ACM Press, New York (2005)
2. Acıiçmez, O., Schindler, W.: A Vulnerability in RSA Implementations due to Instruction Cache Analysis and Its Demonstration on OpenSSL. In: Malkin, T.G. (ed.) CT-RSA 2008. LNCS, vol. 4964, pp. 256–273. Springer, Heidelberg (2008)
3. Archambeau, C., Peeters, E., Standaert, F.-X., Quisquater, J.-J.: Template attacks in Principal Subspaces. In: Goubin, L., Matsui, M. (eds.) CHES 2006. LNCS, vol. 4249, pp. 1–14. Springer, Heidelberg (2006)
4. Brier, E., Clavier, C., Olivier, F.: Correlation Power Analysis with a Leakage model. In: Joye, M., Quisquater, J.-J. (eds.) CHES 2004. LNCS, vol. 3156, pp. 16–29. Springer, Heidelberg (2004)
5. Brumley, D., Boneh, D.: Remote Timing Attacks are Practical. In: Proceedings of the 12th Usenix Security Symposium (2003)
6. Chari, S., Jutla, C.S., Rao, J.R., Rohatgi, P.: Towards Sound Approaches to Counteract Power-Analysis Attacks. In: Wiener, M. (ed.) CRYPTO 1999. LNCS, vol. 1666, pp. 398–412. Springer, Heidelberg (1999)
7. Chari, S., Rao, J.R., Rohatgi, P.: Template attacks. In: Kaliski Jr., B.S., Koç, Ç.K., Paar, C. (eds.) CHES 2002. LNCS, vol. 2523, pp. 13–28. Springer, Heidelberg (2003)
8. Coron, J.-S., Goubin, L.: On Boolean and Arithmetic Masking against Differential Power Analysis. In: Paar, C., Koç, Ç.K. (eds.) CHES 2000. LNCS, vol. 1965, pp. 231–237. Springer, Heidelberg (2000)

9. Dhem, J.-F., Koeune, F., Leroux, P.-A., Mestré, P.-A., Quisquater, J.-J., Willems, J.-L.: A Practical Implementation of the Timing Attack. In: Schneier, B., Quisquater, J.-J. (eds.) CARDIS 1998. LNCS, vol. 1820, pp. 175–191. Springer, Heidelberg (2000)
10. Gierlichs, B., Lemke, K., Paar, C.: Templates vs. Stochastic Methods. In: Goubin, L., Matsui, M. (eds.) CHES 2006. LNCS, vol. 4249, pp. 15–29. Springer, Heidelberg (2006)
11. Gierlichs, B., Batina, L., Tuyls, P., Preneel, B.: Mutual Information Analysis - A Generic Side-Channel Distinguisher. In: Oswald, E., Rohatgi, P. (eds.) CHES 2008. LNCS, vol. 5154, pp. 426–442. Springer, Heidelberg (2008)
12. Huss, S.: Zur interaktiven Optimierung integrierter Schaltungen. Dissertation, TU München, Fakultät für Elektrotechnik und Informationstechnik (1982)
13. Joye, M., Paillier, P., Schoenmakers, B.: On Second-Order Differential Power Analysis. In: Rao, J.R., Sunar, B. (eds.) CHES 2005. LNCS, vol. 3659, pp. 293–308. Springer, Heidelberg (2005)
14. Kardaun, O.J.W.F.: Classical Methods of Statistics. Springer, Berlin (2005)
15. Kasper, M., Schindler, W., Stöttinger, M.: A Stochastic Method for Security Evaluation of Cryptographic FPGA Implementations (to appear in Proc. FPT 2010)
16. Kocher, P.: Timing Attacks on Implementations of Diffie-Hellman, RSA, DSS and Other Systems. In: Koblitz, N. (ed.) CRYPTO 1996. LNCS, vol. 1109, pp. 104–113. Springer, Heidelberg (1996)
17. Kocher, P., Jaffe, J., Jun, B.: Differential Power Analysis. In: Wiener, M. (ed.) CRYPTO 1999. LNCS, vol. 1666, pp. 388–397. Springer, Heidelberg (1999)
18. Lehmann, E.L.: Testing Statistical Hypotheses., 2nd edn. Chapman & Hall, New York (1994) (reprinted)
19. Lemke-Rust, K., Paar, C.: Analyzing Side Channel Leakage of Masked Implementations with Stochastic Methods. In: Biskup, J., López, J. (eds.) ESORICS 2007. LNCS, vol. 4734, pp. 454–468. Springer, Heidelberg (2007)
20. Menezes, A.J., van Oorschot, P.C., Vanstone, S.C.: Handbook of Applied Cryptography. CRC Press, Boca Raton (1997)
21. Messerges, T.S.: Using Second-Order Power Analysis to Attack DPA Resistant Software. In: Paar, C., Koç, Ç.K. (eds.) CHES 2000. LNCS, vol. 1965, pp. 238–251. Springer, Heidelberg (2000)
22. Oswald, E., Mangard, S.: Template Attacks on Masking — Resistance is Futile. In: Abe, M. (ed.) CT-RSA 2007. LNCS, vol. 4377, pp. 243–256. Springer, Heidelberg (2006)
23. Peeters, E., Standaert, F.-X., Donckers, N., Quisquater, J.-J.: Improved Higher-Order Side-Channel Attacks with FPGA Experiments. In: Rao, J.R., Sunar, B. (eds.) CHES 2005. LNCS, vol. 3659, pp. 309–323. Springer, Heidelberg (2005)
24. Schindler, W.: Maße mit Symmetrieeigenschaften. Habilitationsschrift, TU Darmstadt, Fachbereich Mathematik Darmstadt (1998)
25. Schindler, W.: A Timing Attack against RSA with the Chinese Remainder Theorem. In: Paar, C., Koç, Ç.K. (eds.) CHES 2000. LNCS, vol. 1965, pp. 110–125. Springer, Heidelberg (2000)
26. Schindler, W., Koeune, F., Quisquater, J.-J.: Unleashing the Full Power of Timing Attack. Catholic University of Louvain, Technical Report CG-2001/3
27. Schindler, W., Koeune, F., Quisquater, J.-J.: Improving Divide and Conquer Attacks Against Cryptosystems by Better Error Detection / Correction Strategies. In: Honary, B. (ed.) IMA 2001. LNCS, vol. 2260, pp. 245–267. Springer, Heidelberg (2001)

28. Schindler, W.: Optimized Timing Attacks against Public Key Cryptosystems. Statist. Decisions 20, 191–210 (2002)
29. Schindler, W.: A Combined Timing and Power Attack. In: Paillier, P., Naccache, D. (eds.) PKC 2002. LNCS, vol. 2274, pp. 263–279. Springer, Heidelberg (2002)
30. Schindler, W.: Measures with Symmetry Properties. Lecture Notes in Mathematics, vol. 1808. Springer, Berlin (2003)
31. Schindler, W.: On the Optimization of Side-Channel Attacks by Advanced Stochastic Methods. In: Vaudenay, S. (ed.) PKC 2005. LNCS, vol. 3386, pp. 85–103. Springer, Heidelberg (2005)
32. Schindler, W., Lemke, K., Paar, C.: A Stochastic Model for Differential Side Channel Analysis. In: Rao, J.R., Sunar, B. (eds.) CHES 2005. LNCS, vol. 3659, pp. 30–46. Springer, Heidelberg (2005)
33. Schindler, W.: Advanced Stochastic Methods in Side Channel Analysis on Block Ciphers in the Presence of Masking. Math. Crypt. 2, 291–310 (2008)
34. Schramm, K., Paar, C.: Higher Order Masking of the AES. In: Pointcheval, D. (ed.) CT-RSA 2006. LNCS, vol. 3860, pp. 208–225. Springer, Heidelberg (2006)
35. Standaert, F.-X., Koeune, F., Schindler, W.: How to Compare Profiled Side-Channel Attacks. In: Abdalla, M., Pointcheval, D., Fouque, P.-A., Vergnaud, D. (eds.) ACNS 2009. LNCS, vol. 5536, pp. 485–498. Springer, Heidelberg (2009)
36. Standaert, F.-X., Malkin, T.G., Yung, M.: A unified framework for the analysis of side-channel key recovery attacks. In: Joux, A. (ed.) EUROCRYPT 2009. LNCS, vol. 5479, pp. 443–461. Springer, Heidelberg (2010)
37. Waddle, J., Wagner, D.: Towards Efficient Second-Order Analysis. In: Joye, M., Quisquater, J.-J. (eds.) CHES 2004. LNCS, vol. 3156, pp. 1–15. Springer, Heidelberg (2004)
38. Walter, C.D.: Montgomery Exponentiation Needs No Final Subtractions. IEE Electronics Letters 35(21), 1831–1832 (1999)

Survey of Methods to Improve Side-Channel Resistance on Partial Reconfigurable Platforms

Marc Stöttinger[1], Sunil Malipatlolla[2], and Qizhi Tian[1]

[1] Technische Universität Darmstadt
Department of Computer Science
Integrated Circuits and Systems Lab
Hochschulstraße 10
64289 Darmstadt, Germany
{stoettinger,tian}@iss.tu-darmstadt.de

[2] CASED (Center for Advanced Security Research Darmstadt)
Mornewegstraße 32
64289 Darmstadt, Germany
sunil.malipatlolla@cased.de

Abstract. In this survey we introduce a few secure hardware implementation methods for FPGA platforms in the context of side-channel analysis. Side-channel attacks may exploit data-dependent physical leakage to estimate secret parameters like a cryptographic key. In particular, IP-cores for security applications on embedded systems equipped with FPGAs have to be made secure against these attacks. Thus, we discuss how the countermeasures, known from literature, can be applied on FPGA-based systems to improve the side-channel resistance. After introducing the reader to the FPGA technology and the FPGA reconfiguration workflow, we discuss the hiding-based countermeasure against power analysis attacks especially designed for reconfigurable FPGAs.

Keywords: FPGAs, side-channel attacks, countermeasures, power analysis attacks.

1 Introduction

The market for embedded systems has grown rapidly in the last years and the *Static Random Access Memory*-based (SRAM-based) *Field Programmable Gate Arrays* (FPGAs) are becoming increasingly popular as building blocks of such electronic systems. The advantages being easy design modification (reconfigurability), rapid prototyping, economical cost for low volume production, lower startup cost and better maintenance in comparison to fully-customized *Application Specific Integrated Circuits* (ASICs), and availability of sophisticated design and debugging tools. The integration of FPGAs in embedded systems allows to carry out complex and time-consuming operations at moderate costs in terms of power consumption, providing a higher degree of flexibility than ASICs. In the context of mobility and inter-device communication, embedded systems should be secure against attacks on the cryptographic primitives if they support security applications. In particular, secure communication channels between devices

A. Biedermann and H. Gregor Molter (Eds.): Secure Embedded Systems, LNEE 78, pp. 63–84.
springerlink.com

within trusted environments are mandatory nowadays. A Set-Top Box for IP-TV, for instance, has to provide large computational power for streaming high definition videos, and a secure channel has to ensure secure transportation of the streaming content.

The throughput is limited for realization of cryptographic algorithms in software but on the other hand, hardware methods offer high speed and large bandwidth, providing real-time encryption and decryption if needed. ASICs and FPGAs are two distinct alternatives for implementing cryptographic algorithms in hardware.

Today, one of the biggest threats on implementations of cryptographic algorithms in hardware devices are *Side-Channel Attacks* (SCA). In particular, Power Analysis has become an important branch in cryptology since 1999 [13]. With such an attack the secret key (a finite number of ones and zeros) can be recovered on basis of monitoring the power consumption of an active crypto system device. Simply speaking, it is not enough to use a straight forward implementation of a cryptographic algorithm to secure an embedded system. Instead, the algorithm has to be adapted to the hardware platform of the applied system. A lot of countermeasures have been proposed by the industry, mainly the smart-card industry, and different academic research groups to prevent side-channel attacks. Unfortunately most of the contributions focus on smart-cards, software, or ASIC based implementations of crypto systems. With the rising usage of FPGAs for high performance and secure applications it is mandatory to improve the *side-channel resistance* with suitable countermeasure methods and applied techniques.

This contibution is structured as follows: First, we introduce the technology platform of the FPGA and the (partial) reconfiguration workflow to the reader. In the second section we recall the basics of SCA attacks for the readers with no or only a little knowledge in this area. After this, we discuss the different countermeasure methods in general. Before we summarize the work in conclusion, we highlight hiding based countermeasures especially developed for reconfigurable platforms in an independent section.

2 Reconfigurable Hardware

2.1 FPGA Platform Technology

Reconfigurability of a system is the capability to change its behavior i.e., by loading a binary data stream which is called as the *configuration bitstream* in FPGA terminology. One such group of devices offering the reconfigurability feature are FPGAs and others like *Complex Programmable Logic Devices* (CPLDs). The granularity of a reconfigurable logic is the size of the smallest unit that can be addressed by the programming tools. Architectures having finer granularity tend to be more useful for data manipulation at bit level, and in general for combinatorial circuits. On the other hand, blocks with a high granularity are better suited for higher levels of data manipulation, e.g. for developing circuits at register transfer level. The level of granularity has a great impact on the device configuration time. Indeed, devices with low granularity (called as fine-grained devices) for example, FPGAs, require many configuration points producing a bigger vector data for reconfiguration. The extra routing when compared to CPLDs has an unavoidable cost on power and area. On the other hand, devices with high granularity (called as coarse-grained devices) for example, CPLDs, have a tendency of decreased

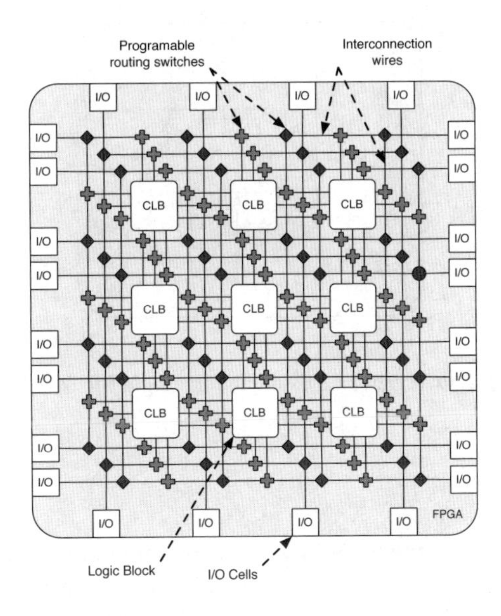

Fig. 1. Architecture of an FPGA

performance when dealing with computations smaller than what their granularity is. In essence, FPGAs can contain large designs while CPLDs can contain small designs only because of fewer number of logic blocks available in the latter devices.

FPGAs are programmable semiconductor devices that are based around a matrix of *Configurable Logic Blocks* (CLBs) connected via programmable interconnects as shown in its general architecture in Figure 1. As opposed to ASICs where the device is custom built for a particular design, FPGAs can be programmed to the desired application or functionality requirements. The CLB is the basic logic unit in an FPGA and exact number of CLBs and their features vary from device to device. In general, a CLB consists of a configurable *Look Up Table* (LUT), usually with 4 or 6 inputs, some selection circuitry (Multiplexer, etc.), and Flip-Flops (FF). The LUT is highly flexible and can be configured to handle combinatorial logic, shift registers or RAM. A high level CLB overview is shown in Figure 2. While the CLB provides the logic capability, flexible interconnect routing, routes the signals between CLBs and to and from Input/Output (I/O) pins.

The design software makes the interconnect routing task hidden to the user unless specified otherwise, thus significantly reducing the design complexity. Current FPGAs provide support for various I/O standards thus providing the required interface bridge in the system under design. I/O in FPGAs is grouped in banks with each bank independently able to support different I/O standards. Today's leading FPGAs provide sufficient I/O banks, to allow flexibility in I/O support. Embedded block RAM memory is available in most FPGAs, which allows for on-chip memory storage in the design. For example, Xilinx FPGAs provide up to 10 Mbits of on-chip memory in 36 kbit blocks that can support true dual-port operation. Also, digital clock management is provided by most FPGAs which acts as the clocking resource for the design.

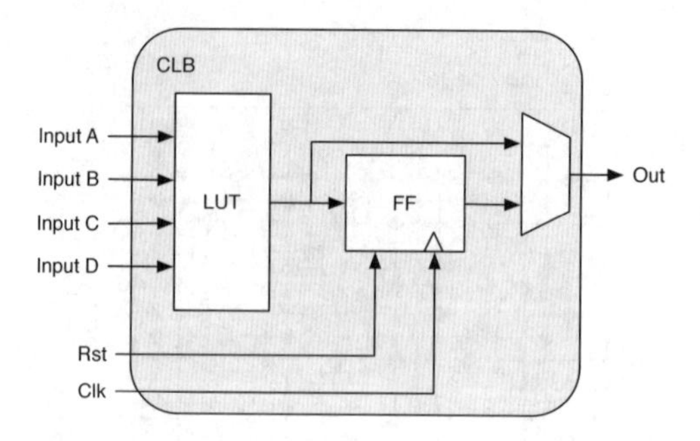

Fig. 2. Overview of a Configurable Logic Block (CLB)

2.2 Terminology

FPGA technology provides the flexibility of on-site programming and re-programming without going through re-fabrication with a modified design. Some of the SRAM-based FPGAs support a special feature called *Partial Reconfiguration* (PR). PR takes the flexibility of FPGAs one step further, allowing modification of an operating design by loading a partial configuration file, usually called as a partial bitfile. After a full bitfile configures the FPGA, a partial bitfile may be downloaded to modify the reconfigurable regions in the FPGA without affecting the applications that are running on those parts of the device that are not being reconfigured. The logic in the FPGA design is divided into two different types, reconfigurable (or dynamic) logic, i.e., the portion being reconfigured and static logic, i.e., the portion resuming the work. The static logic remains continuously functioning and is completely unaffected by the loading of a partial bitfile. Whereas, the configuration in reconfigurable logic is replaced by the partial bitfile. If the configuration of the FPGA is changed at run-time, i.e., the system is neither stopped nor switched off, then it is called as *Dynamic PR* (DPR) and is supported by commercial FPGAs like Xilinx Virtex series. Additionally, if the system triggers the reconfiguration by itself then it is a self-reconfigurable system which does not require the use of internal FPGA infrastructure, but is often assumed.

The area of the FPGA that is reconfigured is called the *Partially Reconfigurable Region* (PRR). A PRR typically consists of a number of CLBs and functional blocks and a design can have more than one PRR. The module to be placed inside the PRR is called a *Partially Reconfigurable Module* (PRM), which is the specific configuration of the PRR, and at least two PRMs are needed per PRR. In many cases, the assignment of PRMs to a PRR is fixed (non-relocatable) though in principle, a PRM may be configured to different PRRs [29]. There are different types of configurations possible in order to achieve PR like "Differential", "Partial", and "Relocatable". The "differential configuration" contains, only difference between current and new configuration whereas, "partial configuration" contains a new configuration part of an FPGA which is independent of the current configuration. In "relocatable configuration", the PRM can be moved to a different location on an FPGA without modification. However, in the following we address only the "partial configuration" technique.

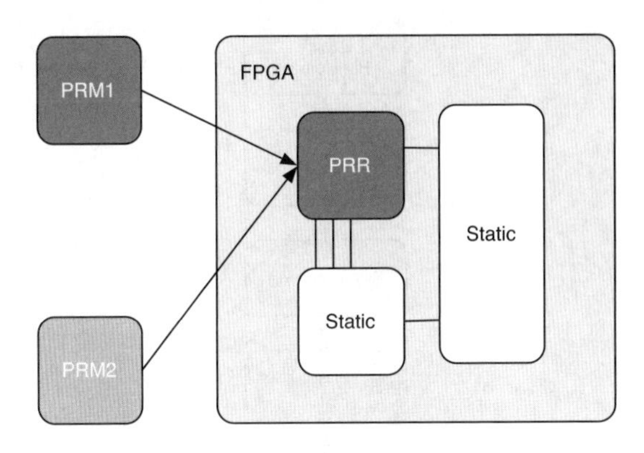

Fig. 3. Partial Reconfiguration in FPGAs

In Figure 3, we see that two PRMs which are mutually exclusive in time will be placed in the PRR inside the FPGA, i.e., only one PRM can be assigned to a given PRR at a given time. The remaining region in the FPGA which is outside the PRR is the static region, where the application which needs to be run uninterruptedly, is placed. In FPGAs supporting dynamic PR, single configuration units can be read, modified and written. For example in Xilinx FPGAs, different Virtex-Series support different PR schemes. Virtex-II/Virtex-II Pro supported initially column-based and later also tile-based PR. Current FPGAs like Virtex-4 and Virtex-5 support tile-based PR only. A minimal reconfigurable unit is one configuration frame which contains 16 x 1 CLBs, 1 clock and 2 IO blocks. The partial reconfiguration of an FPGA is done through the *Internal Configuration Access Port* (ICAP), a built-in hard core *Intellectual Property* (IP) module available on the FPGA. The ICAP module is controlled by the software driver for the processor (e.g., Xilinx's softcore MicroBlaze or IBM's hardcore PowerPC processor) on the FPGA.

A *Macro* is a predefined configuration for one or more FPGA components and a *Bus Macro* is used for communication between regions on an FPGA. In a design containing static and partially reconfigurable regions, the signals between static areas can be routed through reconfigurable areas and the signals between a static and a reconfigurable area are routed through a bus macro. In Virtex-4 devices, each bus macro spans 2 or 4 CLBs. The bus macro has 8-bit signals and is unidirectional, i.e., signals can be routed left-to-right, right-to-left, top-to-bottom or bottom-to-top. A correct macro has to be used, based upon data flow and on which side of the PRR it has to be placed. In Virtex-5 devices, DPR is very similar to Virtex-4 but here the configuration frame is 20 CLBs high and 1 CLB wide. Virtex-5 devices support single slice bus macros which are also unidirectional and can support 4-bit signals. They can be placed anywhere within the DPR area, unlike on the boundary between static and reconfigurable areas in Virtex-4 devices, and provide enable/disable signals to avoid glitches on output signals from a PRR.

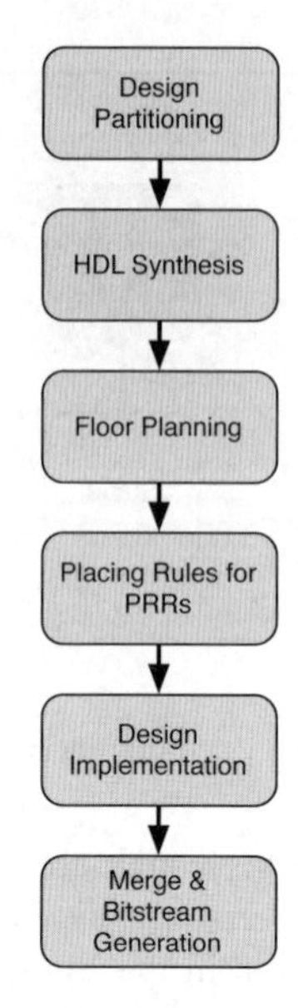

Fig. 4. PR Design Flow

We can draw an analogy between a processor's context switching and the PR of an FPGA. PR is basically a time multiplexing of hardware resources (partial bitfiles) as compared to time multiplexing of software resources (programs or applications) run on a general purpose processor. Many advantages can be gained by such time multiplexing of hardware resources on an FPGA, like reducing the size of an FPGA required for implementing a given function with consequent reductions in cost and/or power consumption. PR provides flexibility in the choices of algorithms or protocols available for an application and it enables new techniques in design security, as explained in section 5. Also, PR improves FPGA fault tolerance and accelerates configurable computing.

2.3 General Workflow for Partial Reconfiguration

Implementing a PR design is similar to implementing multiple non-PR designs that share common logic. In PR, partitions are used to ensure that the common logic between the multiple designs is identical. As a part of the PR design flow, a synthesis of *Hardware Description Language* (HDL) source to netlists for each module is done. The appropriate netlists are implemented in each design to generate the full and partial bitfiles for that configuration of corresponding CLBs and routing paths. The static logic from the previous (first) implementation is shared among all subsequent design implementations. There are multiple reconfigurable modules for each reconfigurable partition in the overall project. The modules are synthesized in a bottom-up fashion, resulting in many netlists associated with each reconfigurable partition. The implementation is then done top-down, which defines a specific set of netlists, called a configuration.

The individual steps in PR design flow as shown in Figure 4, can be explained as follows [11]:

- **Design Partitioning:** A partially reconfigurable FPGA design project is more complex than a normal FPGA design project; therefore a clear file and directory structure will make project management easier. Modules on the top level have to be partitioned into static and reconfigurable components. Each instance of a component on top level is either static or reconfigurable. One reconfigurable instance denotes one PRR in the design and there can be multiple PRMs associated with that PRR. The entity names for PRMs must match component name of PRR instance on top level. The top level contains no logic but only black box instantiations of static and reconfigurable modules including FPGA primitives. Whereas, the modules (static and reconfigurable) themselves contain no FPGA primitives, no bus macros but contain only the actual functionality of the designs. Bus macros are inserted in top level between static components and PRRs.
- **HDL Synthesis:** Top level, static components and PRMs are all synthesized separately. For all components except for top level, the synthesis process option "*Automatic Insertion of I/O Buffers*" should be disabled whereas the option "*Keep Hierarchy*" is enabled for all components including top level. Synthesis of modules can be done using separate ISE projects or using XST scripts or a third party synthesis tools. The results being netlist of the top level with instantiated black box components and netlist of the synthesized logic for other modules.
- **Floor Planning:** A PRR area has to be fixed by constraints, i.e., the required size of PRR area needs to be estimated and then addressed by coordinates of slices from the lower left corner to the upper right. Static logic does not need to be constrained but can be done, following which bus macros have to be placed. Constraints for PRRs and bus macros are defined in a separate file called User Constraints File (UCF).
- **Placing Rules for PRRs:** PRR areas have to span full CLBs with their lower left coordinate being even and the upper right coordinate being odd. If the synthesized design uses functional blocks (e.g. Block RAM) then separate range constraints must be given. Virtex-4 bus macros must cross the boundary between static and reconfigurable area and coordiantes are always even, whereas Virtex-5 uses single slice bus macros which need to be inside PRR.
- **Design Implementation:** The top level netlist is initialized with constraints from the UCF file. PRR locations are reserved and their instances are marked reconfigurable after which bus macros are imported and placed. For the static part, the top level is merged with its components, and then *Map, Place & Route* are done.
- **Merge and bitstream generation:** PRM implementations are merged with static implementations before applying design rule checks and generating a bitfile for the full design. Bitfile for static design with empty PRRs is generated and partial bitfiles are generated as difference between full design and design with holes. Also blanking bitstreams can be generated which are used to erase PRR in order to save some of the hardware resources on the FPGA.

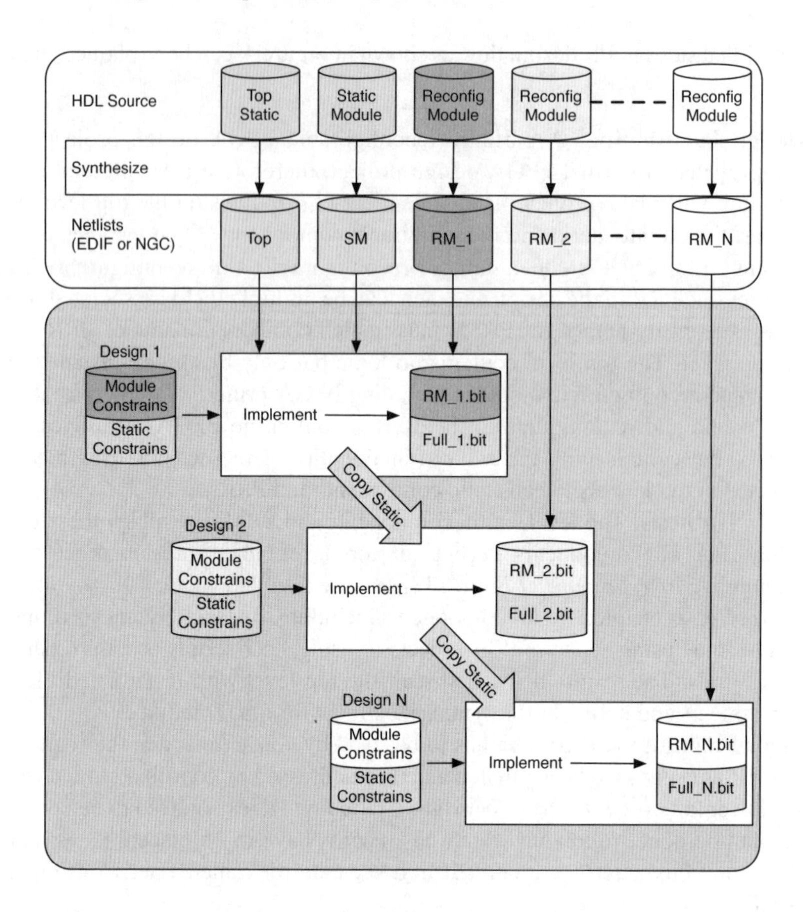

Fig. 5. Software Tool Flow for Partial Reconfiguration in Xilinx FPGAs

2.4 Workflow for PR in Xilinx FPGAs

As an example, the software tool flow for Xilinx PR design flow is shown in Figure 5. Currently, Xilinx FPGAs are the only reconfigurable devices offering the PR feature. Xilinx provides various tools to support each of the steps in the PR design flow. For example, the design partitioning and HDL synthesis can be done using Xilinx's ISE tool. Xilinx's PlanAhead suite can be used for defining PRRs on top level, assignment of PRMs to PRRs, floor planning of PRRs and placement of bus macros. PlanAhead also runs design rule checks and does Map, Place & Route, Merge & Bitfile generation. In order to load the partial bitstreams onto the FPGA, ICAP is used. Xilinx's EDK/XPS tool includes the controller for ICAP, i.e., HWICAP which comes with a software driver for a softcore MicroBlaze processor and is controllable by software. Once the full and partial bitfiles are generated the design can be implemented on the FPGA.

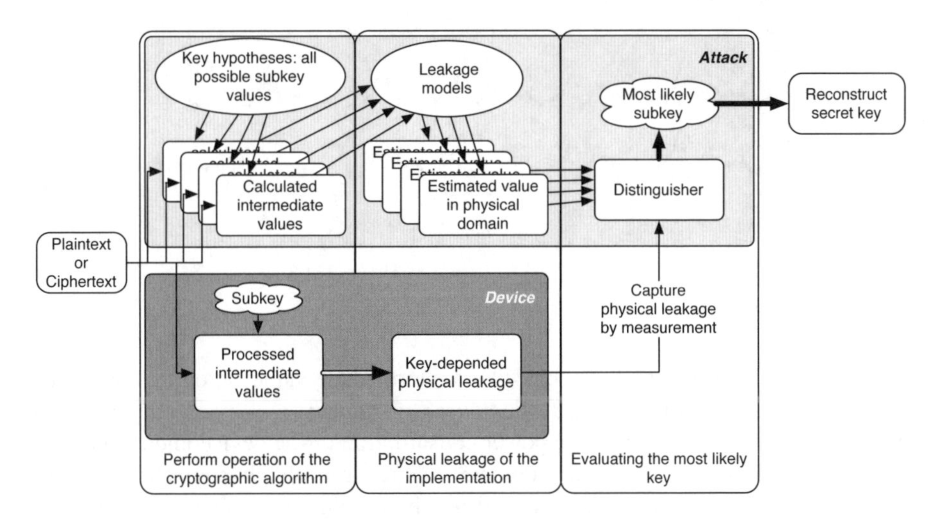

Fig. 6. Workflow of a side-channel attack

3 Side-Channel Attacks

Today, *side-channel analysis* (SCA) attacks are one of the biggest threats to cryptographic implementations in embedded systems. Even mathematically secure algorithms may be susceptible to side-channel methods. SCA exploits the properties of the *implemented* cryptographic algorithm to access information about the processed data and the secrets of the cryptographic device. The different behaviors of the implementations in different physical domains such as time or energy during runtime may generate an exploitable signature. This leaking information, like the processing time, the power consumption, or the electro-magnetic radiation, may be used to gain knowledge about the secrets of the crypto system. The physical leakage of the device transforms the problem of finding the secret private key as a whole into individually finding all the subkeys and then combining them to reconstruct the secret key.

The attacking scenario for a side-channel attack is as follows: First, the adversary needs to have access to the device for a limited time, to measure a physical property of the device, e.g., power or time, during the execution of the cryptographic algorithm by the system. Additionally, he also captures the public input data or output data, in most cases the plaintext or ciphertext, and observes the monitor process of the behavior of the crypto system in the chosen physical domain. Without a deeper knowledge of the implementation, the adversary calculates a subset of intermediate values based on the knowledge, which cryptographic algorithm was processed during the measurement. The intermediate values thereby rely on the additionally collected plaintext or ciphertext during the measurement and a key hypotheses. For each plaintext or ciphertext the attacker gets a set of intermediate values, which equals the number of the key hypothesis. Hereby, the attacker does not consider the complete bit length of the key, instead he focuses on a subkey to reduce the number of hypotheses. In the next phase the attacker estimates the physical measurable values, e.g. power consumption, using

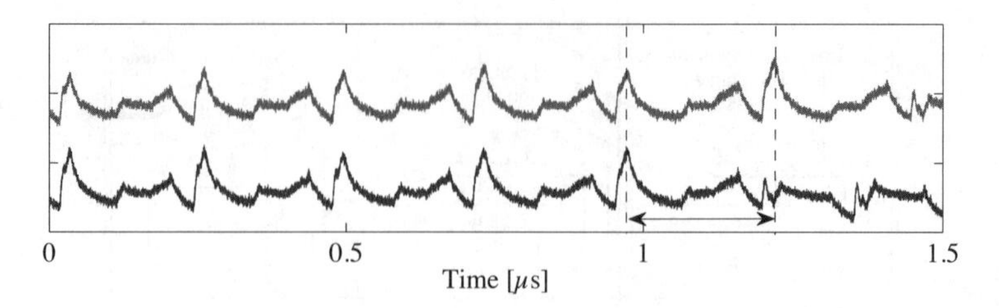

Fig. 7. Power traces from an SPA attack

a leakage model, to transform the calculated intermediate values into the physical domain. In the last step of the attack, the adversary compares with a distinguisher, the measured physical leakage like the power consumption of the device under attack and the estimated values in the physical domain. The result of the distinguisher will point out the subkey which most likely is used for the cryptographic primitive during the monitor procedure of the attack. Figure 6 depicts the generic workflow of an SCA attack described above.

In this contribution, we focus on side-channel attacks which exploit the power consumption of the target device, i.e., we address power analysis attacks. These attacks use only information leaks in the power consumption of a circuit, and no deeper knowledge about the implementation of a cryptographic algorithm is necessarily required. From the literature, the following types of attacks are well-known in the area of power attacks:

- *Simple Power Analysis* (SPA) attack,
- *Differential Power Analysis* (DPA) attack,
- *Template Attack* and
- *Stochastic Approach*

In 1999, Kocher et. al [13] first discovered this most elementary side-channel attack. During an SPA attack, the adversary tries to interpret the captured power trace directly to extract information about the key or the subkey. Especially cryptographic primitives with subkey-value dependent branches in the control flow are very vulnerable to this attack. For instance, SPA attacks on weak RSA and *elliptic curve cryptography* (ECC) implementations are well-known in literature [6], [18], [13]. Based on each bit value of these cryptographic primitives, the intermediate data is processed with different operations. In case of a weak implementation, the processing time of the different operations is significant and thus the key-dependent processing order is directly readable from only one captured power trace. Figure 7 illustrates, how clearly the signature of the trace is key-dependent. These traces were recorded from a MCEliece cryptographic scheme during the decryption [21] process.

The second attack, which Kocher also introduced in 1999, was the differential power analysis attack [13]. Compared with the SPA, the DPA needs more power traces to extract information about the key. In case of the DPA attack, the attacker does not need to distinguish by himself which hypothesis is true for referring to the most likely key or subkey. The method uses statistical techniques to evaluate the hypothesis and thus is more resistant to a low signal-to-noise ratio. Therefore, the attacker has to collect more

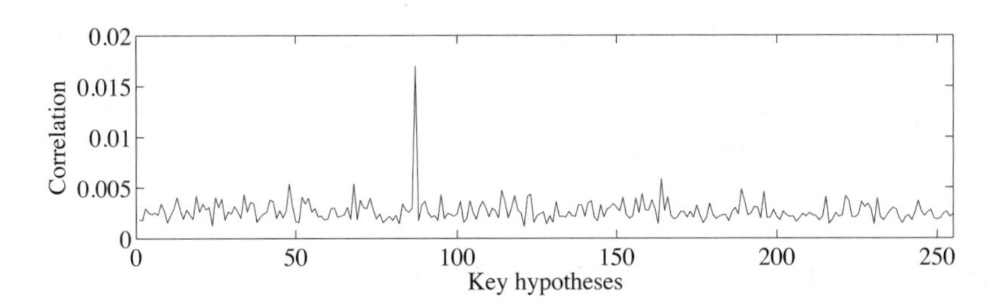

Fig. 8. Correlation result of a DPA attack

power traces of the device during runtime to get the required number of measurement samples for applying the statistical techniques. A variant of the DPA is the *correlation power analysis* (CPA), which uses the *Pearson Correlation Coefficient* for the statistical evaluation. Based on the simple procedure and the quality of the results compared to the effort, the CPA has been established as the most common attack in this area of side-channel attacks. Thus, we will introduce this method a little bit more in detail to the reader.

The "distinguisher" statistically evaluates, how good the estimated hypotheses is, based on matching the estimated power consumption levels to the recorded power consumption levels for several time instants. For an evaluation, the sets of data as mentioned above are denoted as vectors and are sorted accordingly to the key hypotheses. Roughly speaking, each vector of the estimated power consumption levels is based on one key hypothesis and the different input or output data recorded during the monitor procedure of the device. For each key hypothesis, thus, exactly one vector of estimated power consumption levels exists. If one vector of estimated power consumption levels fits better for one time instant than the other vectors with a very high probability to the measured power consumption levels, then the key hypothesis of this vector is with a very high probability the correct guessed key. Figure 8 illustrates an example of a successful DPA attack on a subkey of the block cipher AES. The clearly distinguishable correlation maximum at position 91 on the X-axis marks the most likely value of the used subkey during the measurement phase.

The disadvantage of this method is the need for a higher amount of recorded power traces which may need access to the device for a quite long duration. Depending on the attacking scenario, the attacker may suffer from the disadvantage of the DPA and has not sufficient time to gather enough power traces for a successful DPA attack. In 2002, Rohatghi et al. introduced the Template Attack [5], which handles the disadvantage of the DPA by splitting the attack into two phases: a profiling phase and an attack phase. During the profiling phase the attacker analyses an identical device with full access to all the parameters, for e.g. the keys, to characterize the key-dependent power consumption. The adversary only needs a few measurements of the attacked device to reveal successfully the secret key after characterizing the device, instead of requiring thousands of power measurements for a successful attack. The cost of this efficient attack phase is the expensive preparation in profiling phase on the training device, in which for every subkey, thousands of power traces have to be collected to characterize each possible key value of this focused subkey. However, Schindler proposed in 2005 an optimized approach [20] for attacks which are based on a profiling phase by applying advanced stochastic methods for the subkey characterization. Compared to the Template Attack, the stochastic approach usually needs one to two orders of magnitude less traces in the profiling phase to charaterize a subkey without highly compromising the efficiency during the attack phase.

4 Countermeasures against Power Attacks

In this section, we discuss countermeasures against DPA attacks in general for introducing the concepts. For lowering the success rate of a DPA attack, an intuitive approach would be to decrease the key-dependent physical leakage, which is used to evaluate the different key hypotheses and identifying the most likely subkey. A weak data-dependent or more precisely an almost key-independent power consumption compromises the hypotheses test in the distinguisher, leading to a false result. The two basic concepts known in the literature to decouple the power consumption form the data-dependency or key-dependency are *Masking* and *Hiding*.

Several countermeasures based on these two concepts were introduced to the community over the years. Hereby, the countermeasures can be applied on both levels, the architecture level and the cell level of the implementation. Each layer offers methods and applied techniques to create the recorded power consumption traces in the eyes of the attacker non-deterministic and uncorrelated to the processed key information. In the following section, we introduce these two basic concepts briefly to the reader and state some examples on both application levels.

4.1 Masking

The fundamental idea is to randomize the sensitive and secret (intermediate) values inside the operations of the cryptographic algorithm. This concept benefits from the platform-independency, because this countermeasure is of algebraic nature and can be applied either on architecture or cell level. Randomization is applied on the cryptographic algorithm so that the intermediate values are randomized but the final result of the cryptographic algorithm is the same as the unmasked version of the algorithm. We want to emphasize that the physical leakage characteristic of the implementation is identical to without the countermeasure. The improved security is provided by the countermeasure of the randomized intermediate values, which makes it very hard for an attacker to formulate the corresponding and correct key hypotheses.

The effect of the randomization process can either be used for blinding a secret operand, for instance see [6], or based on the mathematical composition of the random value and the intermediate value, the secret is separated into at least two parts as described in [4]. The cryptographic algorithm is then processed on the parts of the splitted secret. The fragments are individually independent from the intermediate value. Therefore, the effort to formulate corresponding and correct key hypotheses for the attack is increased. After the execution of the cryptographic operations, the two fragments are again united by the inverse function of the mathematical composition. Thus, the strength of the system strongly depends on the injected entropy and thereby the scrambling of the power consumption.

Masking on Architecture Level. The mathematical composition between the random number and the intermediate values in a masking scheme can either be realized with arithmetic operations or with boolean operations. If the arithmetic function is used to create the mask, for instance addition or multiplication, then that masking scheme is dedicated to the group of arithmetic masking. Multiplicative masking is a very suitable countermeasure for point blinding in the cryptographic primitives ECC or RSA, e.g., see [6] and [12]. The utilized mathematical cyclic fields of these asymmetric cryptographic schemes are very much suitable for multiplicative masking because no de-masking is needed to recover the correct result of these algorithms with an applied mask.

When using a XOR-operation or any other boolean operation to composite a randomized intermediate value, the masking scheme is associated to boolean masking. Block ciphers like AES or DES are the main application fields using boolean masking, e.g. see [10] and [3]. As reported in [19], these masking schemes do not suffer from so called zero-value attacks, which exploit the behavior of the masking scheme by processing zero-values, on symmetric masked cryptographic schemes as in the introduced multiplicative based countermeasure in [9] and [28]. For instance, in masking AES, the *AddRoundKey* operation is blinded by a boolean mask and is then splitted into two parts by a modified *SBox* implementation. The most efficient way for a secure version of the SBox, however, would be a BlockRAM-based Look-Up-Table (LUT), but in terms of the resources on an FPGA it costs too many of them.

Masking on Cell Level. The masking scheme on the cell layer focuses on single bits for the randomization. In [27], [8], and [7], different methods to mask AND-gates were discussed to improve the side-channel resistance. The basic principle may be adapted to FPGAs by dedicated initialized LUTs, with the downside being worse CLB-utilization.

4.2 Hiding

In contrast to masking, hiding does not randomize the intermediate values to decouple the key-dependency from the power consumption levels. Instead, hiding techniques try to manipulate the power consumption at several time instants to cover the key-dependent power consumption. Equalizing the consumption levels for every time instant to get a constant power consumption is one possibility to manipulate the power consumption. As much promising this approach sounds, as much difficult it is to realize in general. For equalizing the power consumption to a constant consumption, the design has been specialized for the given implementation platform. The designer has not only to consider a balance between the gate utilization for a constant number of transitions for every input combination, but also he has to consider the captive effects of the connection wires. Thus, the placement of the logic cells as routing of the connection wires is very important to secure a design with the approach of balancing the power consumption.

The second possibility is to randomize the power consumption levels for every point in time to hide the key-dependent power consumption. This randomization method can be applied on the time and amplitude domains to generate an indeterministic power consumption. The randomization in the amplitude domain is very often realized by additional active circuit elements. Thereby, for e.g., the information-leaking cryptographic

Table 1. Known Hiding Techniques against Power Analysis Attacks

Domains	Architecture level	Cell level
time based	shuffling	clockless logic
	dummy operations	delay insertion
amplitude based	random precharge	dual-rail logic
	noise injection	triple-rail logic

operation is duplicated and is fed with false values to produce noise. A randomization in the time domain is, e.g., realized by changing the synchronous control flow of the algorithm. A randomized or indeterministic execution order of operations inside the algorithm is one example to disturb the time resolution of the power trace.

Table 1 is an exemplary, illustrating some countermeasure methods grouped to the domains that manipulate the power consumption. In the following, we stated some related work regarding hiding countermeasures on the two different application layers.

Architecture Level. Applying a hiding countermeasure which manipulates the timing resolution of a cryptographic algorithm may be implemented straight forward. For instance, the control flow is changed by inserting some unnecessary dummy operation to randomly permute the execution order every time the cryptographic algorithm is being executed. The downside of this approach is a performance loss, because the throughput is lowered due to the extra operations. A more promising technique is shuffling the execution order and thereby randomizing the power consumption levels for several time instants. One constraint for this approach is that the shuffled operations have to be independent from each other. However, in 2009 Madlener et al. introduced a shuffling algorithm for a secured multiplication [16], which is able to permute, not completely independent operations. The conducted experiments on an FPGA platform and a detailed algorithmic description of the proposed secured multiplication scheme (called eMSK) is additionally listed in [23].

Another method to mainpulate the power consumption is to compromise the amplitude domain, e.g., duplicated circuit elements, which operate on uncorrelated values to increase the noise in the amplitude domain is either parallelism or pipelining. In [22] Standaert et al. investigated deterministic pipelining as a countermeasure for AES on a FPGA platform and derived from their experiment that it is not an efficient countermeasure. A more efficient method is to precharge the register elements to compromise the bit transitions, which directly effects the data-dependent power consumption. The secured eMSK in [16] also applies this concept within its countermeasure.

Cell Level. A very efficient method to manipulate the time domain on cell layer is proposed by Lu et al. [15]. This countermeasure injects random delays to change the time instance when the leaking logic is active. With different propagation delay between the registers, the transition activity is moved inside the clock cycle on the FPGA implementation. A manipulation technique, working in the amplitude domain is *dual-rail logtic*. The basic idea behind dual-rail logic designs is to mirror the circuit design in such a way

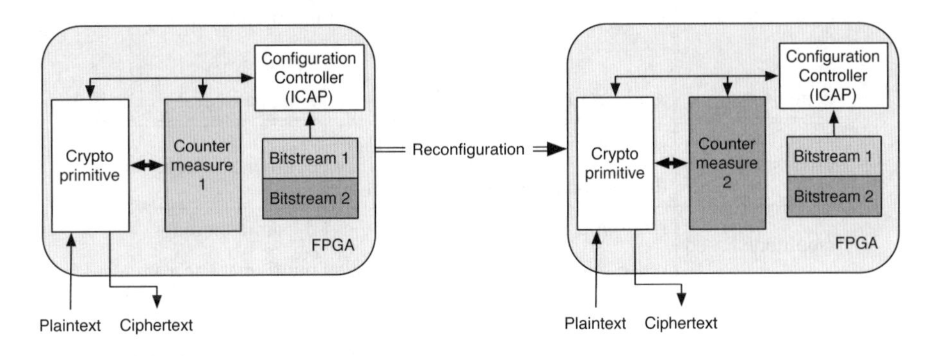

Fig. 9. Basic concept using partial reconfiguration for hiding purpose

that the power consumption is the same for every point in time. Especially for FPGAs, it is difficult to implement secured dual-rail logic based design [24,25,26] to manipulate the amplitude domain by equalizing the power consumption over the complete cryptographic process. Yu et al. developed a Place and Route tool for FPGAs [30], which applies special constraints to assure that the power consumption is equalized. Lomné et al. introduced triple-rail logic [14] to avoid glitches. They also verified their method with conducted experiments on an FPGA platform.

5 Reconfigurable Technology-Based Countermeasure

In this section, we want to discuss methods and concepts to extend the application layers for hiding based countermeasures, especially for reconfigurable platforms. The partial reconfiguration properties of an FPGA offers a designer additional methods to implement hiding schemes to randomize the power consumption. As introduced in section 2, the reconfiguration techniques for FPGAs are quite flexible and offer the designer a broad range of choices to adapt his design during runtime.

The ability to reconfigure the implemented architecture during runtime serves a designer with an advantage to efficiently compare the countermeasures in one design with an acceptable resource utilization. Figure 9 illustrates the fundamental idea of how the partial reconfiguration is used to improve the side-channel resistance of a given design.

Additionally, the combination has not to be a composition of countermeasures, instead the countermeasures may vary during the execution time. The changes of countermeasures can be seen as a shuffling of the leakage characterization, considering the assumption that no countermeasure completely seals the exploitable information leakage of a design. Therefore, the partial reconfiguration based countermeasures randomize the might-be exploitable leakage information of the countermeasure, to increase the effort of the side-channel attack. In case of adapting the countermeasure during runtime, to prevent a small but still explicit leakage, a random number generator might improve this with a random selection of the countermeasure for each execution or at least a small number of executions of the cryptographic primitive.

As stated before in previous section 4.2, the hiding-based countermeasures are not platform-independent. With an adaptive reconfigurable platforms like FPGAs, we can

Table 2. Hiding countermeasures for reconfigurable architectures

Domains	Architecture level	Cell level	Data path level
time based	shuffling dummy operations	clockless logic delay insertion	configurable pipelining
amplitude based	random precharge noise injection	dual-rail logic triple-rail logic	dynamic binding multi-characteristic

extend Table 1 with an additional application layer for hiding countermeasure concepts. Table 2 depicts the novel application layer for hiding countermeasures. The introduced reconfiguration-based hiding countermeasures on the data path level work also on both domains, time and amplitude, to randomize the power consumption.

The hiding countermeasures on the data path level utilizes architectural structures and implementation behavior of different logic blocks, which realize the cryptographic operation for a cryptographic primitive. Especially the exchange of implementation parts via reconfiguring the logic blocks on the FPGA is more effective than using only rerouting and activating already fixed instantiated modules to avoid information leakage based on the connection wiring.

Remark: Besides the adaptation of the countermeasure during runtime and the improvement of the security by the variants of the indeterministic power consumption levels at several time instants, the reconfiguration can also be used to "update" a countermeasure scheme easily, during runtime. Thus, the design might evolve to overcome new side-channel attack schemes without a complete redesign of the implementation by using a modular structure with dedicated areas of countermeasures to strengthen the system.

5.1 Mutating the Data Path

The fundamental idea behind hiding countermeasures on data path level, is to manipulate not the complete architecture at once but concurrently to change only part-wise and independently the behavior of some logic components in the amplitude and time domains. With this approach, the implementation costs and efforts are less than on the cell level but the flexibility and the variation of the distribution is higher than on the architecture level. Based on the modularized randomization, we gain additional combinations to hide the leakage and more variance of the power consumption levels at one time instant.

In the following, we want to introduce to three countermeasures as an example how the new application level can be used to harden a design against power analysis attacks. We discuss the three hiding countermeasures on the data path level by realized example cryptographic primitives.

Configurable Pipelining. To the best knowledge of the authors, the first known countermeasure, which utilizes the reconfiguration techniques of a FPGA was done by

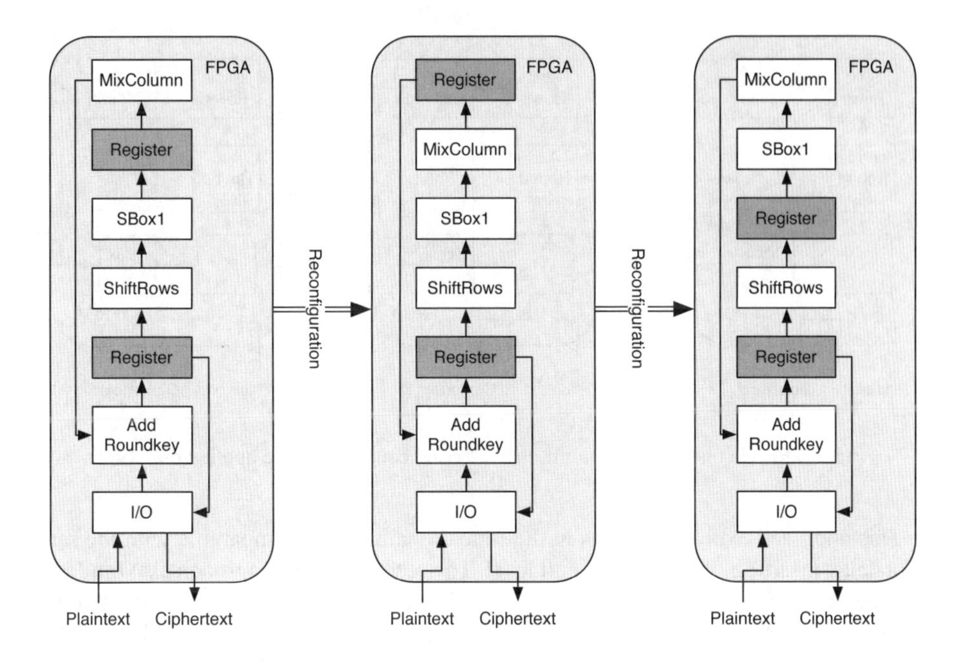

Fig. 10. Dynamic pipelining approach form [17]

Mentes et al. in 2008 [17]. Their approach focuses on a method which variates the pipelining structure and the position of the pipeline stages of a recursive AES implementation. Figure 10 depicts the principle of this countermeasure and how the reconfiguration is used to change the structure.

The difference between the three data path structures is the position of the pipeline register and thereby the number of operations executed within one clock cycle. By changing the pipelining stage for each execution, the attacker is not able, or at least only with a lot more effort, to compare the different power consumption levels at several time instants. The adversary can not determine in which clock cycle the focus and leaking cryptographic operation is executed, thus, the attack success rate will be lowered.

The partial reconfiguration in this countermeasure technique is used to reconnect the input and the output of each cryptographic operation module in the AES, for e.g., the SBox operation, with each other or one of two pipeline registers. The architecture consists only of one partial reconfigurable module. This module interconnects the register and the cryptographic operation modules. Such a reconfigurable interconnect module consists only of wires and no additional switching logic. The additional logic will increase propagation time and thus lowers the throughput. An additional switching logic would also cause a higher power consumption, which might leak out information of the chosen pipeline structure. The configuration time is minimized by just reconfiguring one module, which only consists of static connections instead of reconfiguring the complete AES scheme.

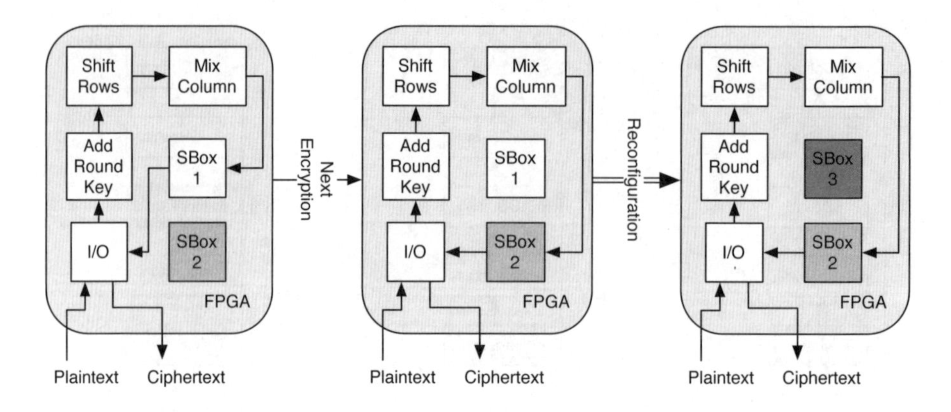

Fig. 11. Basic concept of the multi-characteristic countermeasure applied on AES

This approach explicitly works in the time domain and is comparable with the random delay insertion of [15] on the cell level. The reconfigurable approach also randomizes the delay by selecting the pipelining states and structure via a random number but it does not suffer from a lower throughput because of deeper combinatorial logic paths like in [15]. The downside of the approach of Mentes et al. is the additional time to reconfigure before executing AES once.

Multi-Characteristic of Power Consumption. In 2003, Benini et al. proposed the first countermeasure against DPA attacks working on data path level. The proposed concepts in [1,2] were not directly designed for the reconfigurable platforms, but they are very suitable for FPGAs. In their proposal, they introduced the concept of mutating a data path by using modules with different power consumptions but with the same functionality. The selection of a module is done by a random number generator. The basic idea behind this approach is to change the proportional ratio between the transition activity and the power consumption level. Shuffling the execution modules with different power consumptions and thereby averaging the power consumption of different input data will tamper the correlation between the real and the estimated power consumption levels.

The researchers demonstrated the concept by randomly switching between two different implementations of two arithmetic logic units. They analyzed different dynamic power consumptions but not with a detailed DPA attack. As second application example they discussed a RSA trace with and without this countermeasure, while the runtime in both designs differs. In this example, the RSA power traces seem to be more randomized and not deterministic as the recorded trace of the unsecured version.

We adapted this concept to reconfigurable platforms to improve the strength of the approach by Benini et al. and applied it on an AES scheme. Figure 11 depicts the adapted version applied on an AES scheme. We chose to several different impleneta-tions of the SBox operation because this module has the highest dynamic power consumption according to its complexity.

Instead of instantiating four fixed SBoxes on the FPGA we use the partial reconfiguration technique to reconfigure one of the two modules in the design. The advantage

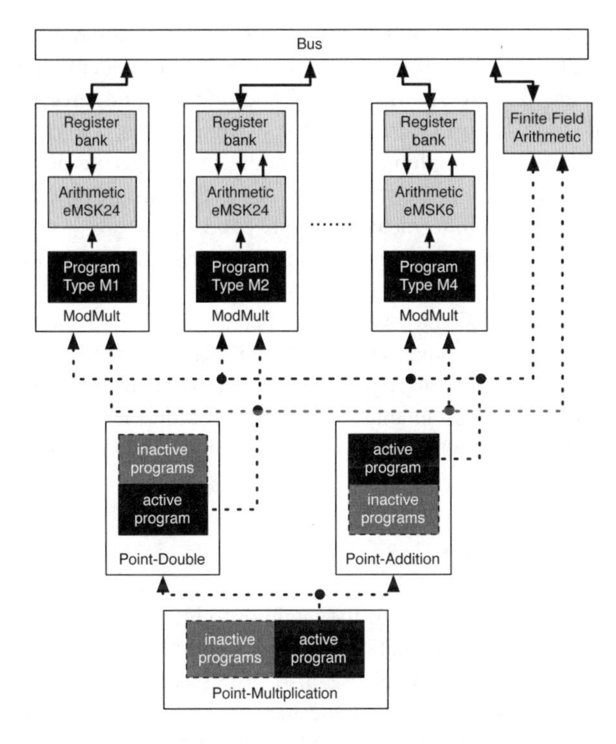

Fig. 12. Concept of dynamic binding applied on the EC-arithmetic layer

is to hide the reconfiguration time by using the second SBox in the system and reduce the complexity of the complete architecture by switching only between two SBoxes. Additionally, this reduces the length of the combinatorial path and the throughput is not as bad as using four fixed implemented SBoxes. Compared to the approach from Mentes et al. in [17], our countermeasure needs more time for the reconfiguration of one SBox, but with the approach of parallel instantiated SBoxes the availability of the cryptographic scheme is not influenced by that.

Dynamic Binding. A combination of the previous reconfigurable countermeasures was proposed in 2009 by Stöttinger et al. [23]. They applied a reconfigurable hiding concept to improve the side-channel resistance of ECC to avoid either SPA attacks or DPA attacks. Weak implementations of ECC are attackable based on the different execution durations of the operations inside ECC. The adaptive hardening technique is also applied on the abstract data path level of the design. This countermeasure, either varies the start and the execution time of the different operations inside ECC and disturbs the signal-to-noise ratio by changing the degree of parallelism of the modular multiplication and squaring operations.

The basic concept of this hardening technique is to dynamically bind a different number of multipliers for executing a Point-Addition or a Point-Doubling. Additionally, executing a squaring operation, the binding has even more degrees of freedom, because this operation can be executed on one of the multiplier modules or on a special squaring module, also with different power consumption and execution duration properties. The usage of eMSK based multiplier [16] additionally randomizes each concurrent multiplication process. To further improve the strength of the system, each eMSK can be exchanged with another implementation model to change the power consumption characterization additionally to the shuffling process. So, this proposed countermeasure unites the strength of the countermeasures in [17] and [1].

Figure 12 depicts the basic structure of the countermeasure. Each EC-operation is implemented as an independent module to modularize the structure and increase the possible combinations of different execution procedures. The control unit for the Point-Doubling, the Point-Addition, and the Point-Multiplication is separated in an active and an inactive part, like the previously proposed countermeasure for multi-characteristic power consumption. This also enables the ability to hide the needed reconfiguration time by reconfiguring the execution program of one module while an EC-operation is processed.

6 Summary and Conclusion

In this survey, we gave an overview of the methods to improve side-channel security of cryptographic primitive implementations on FPGAs. Besides a report of some related work on countermeasures in general, we also discussed their adaptation on FPGAs. Especially, because the hiding-based countermeasures are platform-dependent, it is difficult to transfer hiding techniques from ASICs, like those used in smart-card chips, to FPGAs. To emphasize the potential of the FPGA platform for hiding techniques, we further introduced the reader to the actual reconfiguration design flow. Afterwards, we discussed the first reconfigurable based hiding countermeasures for FPGA platforms and showed how they improve the side-channel resistance on another abstract application layer. Some of these discussed countermeasures can handle the needed configuration time to avoid a lower throughput compared to a non-reconfigurable system.

Acknowledgment

This work was supported by DFG grant HU 620/12, as part of the Priority Program 1148, in cooperation with CASED (http://www.cased.de).

References

1. Benini, L., Macii, A., Macii, E., Omerbegovic, E., Poncino, M., Pro, F.: A novel architecture for power maskable arithmetic units. In: ACM Great Lakes Symposium on VLSI, pp. 136–140. ACM, New York (2003)
2. Benini, L., Macii, A., Macii, E., Omerbegovic, E., Pro, F., Poncino, M.: Energy-aware design techniques for differential power analysis protection. In: DAC, pp. 36–41. ACM, New York (2003)
3. Canright, D., Batina, L.: A very compact "perfectly masked" s-box for aes. In: Bellovin, S.M., Gennaro, R., Keromytis, A.D., Yung, M. (eds.) ACNS 2008. LNCS, vol. 5037, pp. 446–459. Springer, Heidelberg (2008)

4. Chari, S., Jutla, C.S., Rao, J.R., Rohatgi, P.: Towards sound approaches to counteract power-analysis attacks. In: Wiener, M. (ed.) CRYPTO 1999. LNCS, vol. 1666, pp. 398–412. Springer, Heidelberg (1999)
5. Chari, S., Rao, J.R., Rohatgi, P.: Template attacks. In: Kaliski Jr., B.S., Koç, Ç.K., Paar, C. (eds.) CHES 2002. LNCS, vol. 2523, pp. 13–28. Springer, Heidelberg (2003)
6. Coron, J.S.: Resistance against differential power analysis for elliptic curve cryptosystems. In: Koç, Ç.K., Paar, C. (eds.) CHES 1999. LNCS, vol. 1717, pp. 292–302. Springer, Heidelberg (1999)
7. Fischer, W., Gammel, B.M.: Masking at gate level in the presence of glitches. In: Rao, J.R., Sunar, B. (eds.) CHES 2005. LNCS, vol. 3659, pp. 187–200. Springer, Heidelberg (2005)
8. Golic, J.D., Menicocci, r.: Universal masking on logic gate level. IEEE Electronic Letters 40(9), 526–528 (2004)
9. Golic, J.D., Tymen, C.: Multiplicative masking and power analysis of aes. In: Kaliski Jr., B.S., Koç, Ç.K., Paar, C. (eds.) CHES 2002. LNCS, vol. 2523, pp. 198–212. Springer, Heidelberg (2003)
10. Herbst, C., Oswald, E., Mangard, S.: An aes smart card implementation resistant to power analysis attacks. In: Zhou, J., Yung, M., Bao, F. (eds.) ACNS 2006. LNCS, vol. 3989, pp. 239–252. Springer, Heidelberg (2006)
11. Herrholz, A.: Invited Talk on Dynamic Partial Reconfiguration Using Xilinx Tools and Methodologies (February 2010)
12. Joye, M., Tymen, C.: Protections against differential analysis for elliptic curve cryptography. In: Koç, Ç.K., Naccache, D., Paar, C. (eds.) CHES 2001. LNCS, vol. 2162, pp. 377–390. Springer, Heidelberg (2001)
13. Kocher, P.C., Jaffe, J., Jun, B.: Differential power analysis. In: Wiener, M. (ed.) CRYPTO 1999. LNCS, vol. 1666, pp. 388–397. Springer, Heidelberg (1999)
14. Lomné, V., Maurine, P., Torres, L., Robert, M., Soares, R., Calazans, N.: Evaluation on fpga of triple rail logic robustness against dpa and dema. In: DATE, pp. 634–639. IEEE, Los Alamitos (2009)
15. Lu, Y., O'Neill, M.: Fpga implementation and analysis of random delay insertion countermeasure against dpa. In: IEEE International Conference on Field-Programmable Technology (FPT) (2008)
16. Madlener, F., Stöttinger, M., Huss, S.A.: Novel hardening techniques against differential power analysis for multiplication in gf(2n). In: IEEE International Conference on Field-Programmable Technology (FPT) (December 2009)
17. Mentens, N., Gierlichs, B., Verbauwhede, I.: Power and fault analysis resistance in hardware through dynamic reconfiguration. In: Oswald, E., Rohatgi, P. (eds.) CHES 2008. LNCS, vol. 5154, pp. 346–362. Springer, Heidelberg (2008)
18. Örs, S.B., Oswald, E., Preneel, B.: Power-analysis attacks on an fpga - first experimental results. In: Walter, C.D., Koç, Ç.K., Paar, C. (eds.) CHES 2003. LNCS, vol. 2779, pp. 35–50. Springer, Heidelberg (2003)
19. Oswald, E., Mangard, S., Pramstaller, N., Rijmen, V.: A side-channel analysis resistant description of the aes s-box. In: Gilbert, H., Handschuh, H. (eds.) FSE 2005. LNCS, vol. 3557, pp. 413–423. Springer, Heidelberg (2005)
20. Schindler, W.: On the optimization of side-channel attacks by advanced stochastic methods. In: Vaudenay, S. (ed.) PKC 2005. LNCS, vol. 3386, pp. 85–103. Springer, Heidelberg (2005)
21. Shoufan, A., Wink, T., Molter, H.G., Huss, S.A., Kohnert, E.: A novel cryptoprocessor architecture for the mceliece public-key cryptosystem. IEEE Transactions on Computers 99 (2010)(preprints)
22. Standaert, F.X., Örs, S.B., Preneel, B.: Power analysis of an fpga: Implementation of rijndael: Is pipelining a dpa countermeasure? In: Joye, M., Quisquater, J.-J. (eds.) CHES 2004. LNCS, vol. 3156, pp. 30–44. Springer, Heidelberg (2004)

23. Stöttinger, M., Madlener, F., Huss, S.A.: Procedures for securing ecc implementations against differential power analysis using reconfigurable architectures. In: Platzner, M., Teich, J., Wehn, N. (eds.) Dynamically Reconfigurable Systems - Architectures, Design Methods and Applications, pp. 305–321. Springer, Heidelberg (2009)
24. Tiri, K., Verbauwhede, I.: A logic level design methodology for a secure dpa resistant asic or fpga implementation. In: DATE, pp. 246–251. IEEE Computer Society, Los Alamitos (2004)
25. Tiri, K., Verbauwhede, I.: Secure logic synthesis. In: Becker, J., Platzner, M., Vernalde, S. (eds.) FPL 2004. LNCS, vol. 3203, pp. 1052–1056. Springer, Heidelberg (2004)
26. Tiri, K., Verbauwhede, I.: Synthesis of secure fpga implementations. Cryptology ePrint Archive, Report 2004/068 (2004), http://eprint.iacr.org/
27. Trichina, E., Korkishko, T., Lee, K.H.: Small size, low power, side channel-immune aes coprocessor: Design and synthesis results. In: Dobbertin, H., Rijmen, V., Sowa, A. (eds.) AES 2005. LNCS, vol. 3373, pp. 113–127. Springer, Heidelberg (2005)
28. Trichina, E., Seta, D.D., Germani, L.: Simplified adaptive multiplicative masking for aes. In: Kaliski Jr., B.S., Koç, Ç.K., Paar, C. (eds.) CHES 2002. LNCS, vol. 2523, pp. 187–197. Springer, Heidelberg (2003)
29. Xilinx Corporation: Partial Reconfiguration User Guide (September 2009)
30. Yu, P., Schaumont, P.: Secure fpga circuits using controlled placement and routing. In: Ha, S., Choi, K., Dutt, N.D., Teich, J. (eds.) CODES+ISSS, pp. 45–50. ACM, New York (2007)

Multicast Rekeying: Performance Evaluation

Abdulhadi Shoufan and Tolga Arul

CASED (Center for Advanced Security Research Darmstadt)
Mornewegstraße 32
64289 Darmstadt, Germany
{abdul.shoufan,tolga.arul}@cased.de

Abstract. This paper presents a new approach for performance evaluation of rekeying algorithms. New system metrics related to rekeying performance are defined: Rekeying Quality of Service and Rekeying Access Control. These metrics are estimated in relation to both group size and group dynamics. A simultor prototype demonstrates the merit of this unified assessment method by means of a comprehensive case study.

Keywords: Group Rekeying Algorithms, Performance Evaluation, Benchmarking, Simulation.

1 Introduction

While multicast is an efficient solution for group communication over the Internet, it raises a key management problem when data encryption is desired. This problem originates from the fact that the group key used to encrypt data is shared between many members, which demands the update of this key every time a member leaves the group or a new one joins it. The process of updating and distribution of the group key, denoted as group rekeying, ensures forward access control regarding leaving members and backward access control concerning the joining ones. Figure 1 represents a Pay-TV environment as an example for a multicast scenario. A video provider (VP) utilizes a video server (VS) to deliver video content encrypted with a group key k_g. A registration and authentication server (RAS) manages the group and performs group rekeying. Every registered member gets an identity key k_d (e.g., k_0 to k_3 in Figure 1) and the group key k_g. To disjoin member m_2, for instance, the RAS generates a new group key and encrypts it with each of the identity keys of the remaining members. In other words, disjoining a member from a group having n participants costs a total of $n-1$ encryptions on the server side. Obviously, a scalability problem arises.

This problem has been amply addressed in the last decade to reduce the rekeying costs and to achieve a scalable group key management. A large variety of architectures, protocols, and algorithms have been proposed in literature, see [1] to [14]. Although many approaches have been proposed, the reader of related publications lacks a way to compare the results of these solutions to each other. This is attributed mainly to vastly different ways of estimating rekeying costs by

A. Biedermann and H. Gregor Molter (Eds.): Secure Embedded Systems, LNEE 78, pp. 85–104.
springerlink.com

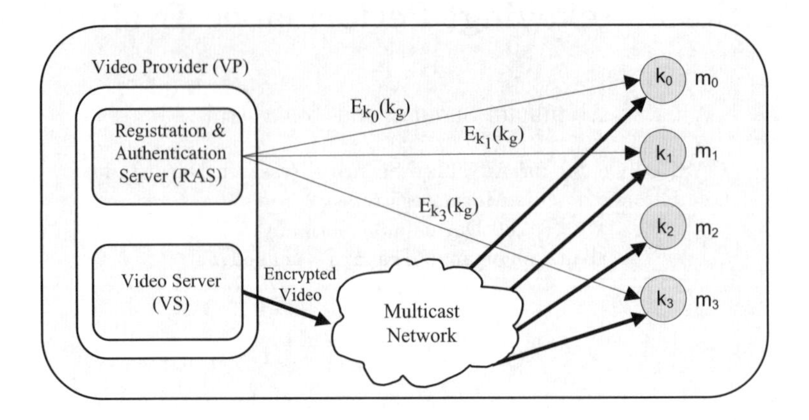

Fig. 1. Pay-TV a potential scenario for secure multicast

different researchers and to the application of highly diverse metrics to express these costs.

In this paper a *rekeying benchmark* is presented, which allows for a *reliable rekeying performance evaluation* and for fair comparison between different rekeying algorithms. This originates from evaluating the rekeying performance on a high abstraction level: Rekeying costs are determined on the system level independent of the evaluated rekeying algorithms themselves and regardless of the underlying cryptographic primitives and execution platform. The rekeying benchmark is realized by a simulator, which supports the execution of different rekeying algorithms with uniform simulation parameters. The simulator estimates unified cost metrics and presents simulation results in the same diagram for comparison.

The rekeying benchmark considers only the costs of cryptographic operations required for rekeying on the server side, which dominate the total cost in most cases. Including other factors is part of future work.

The rest of the paper is organized as follows. Section 2 represents the design concept of the proposed rekeying benchmark. Section 3 describes the realization of this concept as a simulation environment. Section 4 details the benchmark design. To show the advantage of this solution Section 5 provides a case study.

2 Rekeying Benchmark Design Concept

The difficulty of evaluating different rekeying algorithms is attributed to the following three points in the literature:

1. Non-unified performance estimation methods.
2. Non-unified consideration of the input quantities affecting the performance.
3. Non-unified definition of output metrics representing the performance.

2.1 Benchmark Abstraction Model

Rekeying is a solution for group key management in secure multicast. As an essential step in the process of joining and removing members, rekeying performance directly influences the efficiency of this process with major effects on the system behavior. The faster a member can be removed the higher is the system security. The faster a member can be joined the higher is the system quality of service. The performance of a rekeying algorithm directly affects the supportable group size and dynamics. Accordingly, the importance of rekeying performance results from its significance for the system behavior with respect to the following characteristics:

1. Amount of *quality of service* that can be offered to a joining member.
2. Amount of *security* against a removed member.
3. *Scalability* in terms of supportable group sizes.
4. *Group dynamics* in terms of maximal supportable join and disjoin rates.

These characteristics allow for evaluating rekeying performance on a high abstraction level, see Figure 2. To enable a reliable performance evaluation of rekeying algorithms, metrics should be used, which are independent of these algorithms. Therefore, the performance evaluation is settled on the *Benchmark Layer*, which is sperated from *Rekeying Layer* in Figure 2. The introduction of the *Cryptography Layer* and the *Platform Layer* is justified as follows. The Rekeying Layer performs join and disjoin requests based on cryptographic operations such as encryption and digital signing. For each cryptographic primitive a vast selection is available. Taking symmetric-key encryption as an example, rekeying may employ 3DES, AES, IDEA or other algorithms. The same rekeying algorithm behaves differently according to the utilized cryptographic primitives. Further more, the same cryptographic primitive features different performance according to the platform it runs on. This fact remains, even if public-domain libraries such as CryptoLib [17] are utilized to realize cryptographic functions. Consequently, a reliable rekeying benchmark does not only rely on an abstraction from the details of the analyzed rekeying algorithms. Rekeying itself must be decoupled from the underlying cryptographic primitives and from the executing platform.

The abstraction model of Figure 2 introduces essential design aspects for the benchmark:

1. The separation of rekeying algorithms from the cryptographic layer and from the execution platform leads to a substantial acceleration of the evaluation process. This gain is based on the fact that rekeying algorithms to be evaluated do not need to execute cryptographic algorithms anymore. Instead, they just provide information on the required numbers of these operations. The actual rekeying costs are then determined by the benchmark with the aid of timing parameters relating to the used primitives and the execution platform. This point will be detailed in the Section 2.2.

2. From the last point it is obvious that the demand for a reliable rekeying benchmark can not be fulfilled by real-time measurements on prototypes or final products, since these measurements can not be performed independently of the cryptographic primitives and the platform. Instead, for rekeying algorithms to be evaluated fairly and efficiently, some kind of simulation has to be employed.

2.2 Benchmark Data Flow

A good understanding of the benchmark abstraction model can be achieved by investigating the data exchange between its different layers as depicted in Figure 2:

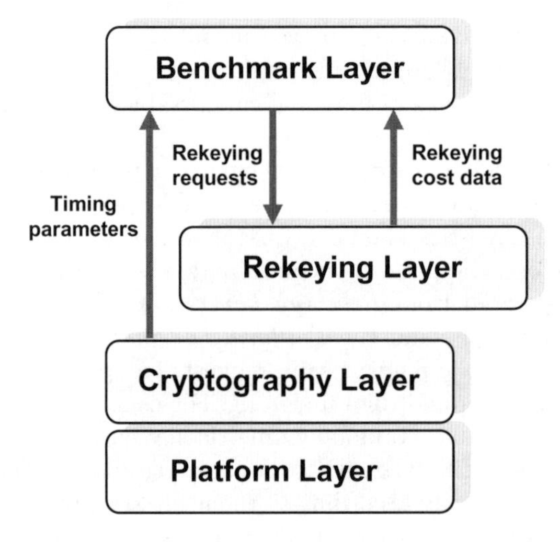

Fig. 2. Data exchange in the rekeying benchmark

1. The rekeying layer receives rekeying requests and executes pseudo rekeying, which means that rekeying algorithms only decide on the cryptographic operations needed for these requests without executing them. This issue is illustrated by the gap between the rekeying and the cryptography layers.
2. The rekeying requests are delivered without any timing information. This means that the rekeying layer is not informed about the temporal distribution of the rekeying requests. This task is assigned to the benchmark layer.
3. The rekeying cost data provide information on the number of the needed cryptographic operations for each rekeying request or request batch.
4. The timing parameters hide the cryptographic primitives and the executing platform to provide a unified cost estimation, which can be used by the benchmark layer for all rekeying algorithms in the same way.
5. To determine the time needed for executing a rekeying request, the benchmark sums up the products of the rekeying cost data and the corresponding timing parameters.

3 Rekeying Benchmark as a Simulation Environment

3.1 Cost Metrics and Group Parameters

Definition 1. ***Required Join Time*** T_J^{sys}
A Required Join Time is a system parameter representing the maximal allowable rekeying time needed to join a member.

Definition 2. ***Actual Join Time*** T_J
An Actual Join Time specifies a join request and is defined as the sum of the waiting time W_J of the join request in the system queue and the rekeying time RT_J consumed by a rekeying algorithm to grant this request:

$$T_J = W_J + RT_J \tag{1}$$

Definition 3. ***Rekeying Quality of Service*** $RQoS$
Rekeying Quality of Service specifies a join request and is defined as the difference between the required join time and the actual join time of this request:

$$RQoS = T_J^{sys} - T_J \tag{2}$$

Definition 4. ***Required Disjoin Time*** T_D^{sys}
A Required Disjoin Time is a system parameter representing the maximal allowable rekeying time needed to disjoin a member.

Definition 5. ***Actual Disjoin Time*** T_D
An Actual Disjoin Time specifies a disjoin request and is defined similarly to T_J as follows:

$$T_D = W_D + RT_D \tag{3}$$

Definition 6. ***Rekeying Access Control*** RAC
Rekeying Access Control specifies a disjoin request and is defined as the difference between the required disjoin time and the actual disjoin time of this request:

$$RAC = T_D^{sys} - T_D \tag{4}$$

Definition 7. ***Maximal Group Size*** n_{max}
Maximal Group Size represents the supportable group size without deterioration of the system requirements of QoS and access control.

Definition 8. ***Maximal Join/Disjoin rates*** λ_{max}/μ_{max}
Maximal Join/Disjoin rates represent the maximal group dynamics which can be supported without deterioration of the system requirements of QoS and access control.

3.2 Evaluation Criteria and Simulation Modes

Depending on the proposed metrics and parameters, rekeying algorithms may be evaluated by checking the following criteria:

1. For a rekeying algorithm to function correctly, it must feature $RQoS$ and RAC values, which are equal to or greater than zero.
2. Considering two rekeying algorithms, which satisfy criterion 1, the algorithm that supports a higher n_{max} features a higher scalability.
3. Considering two rekeying algorithms, which fulfill criterion 1, the algorithm that supports higher λ_{max} or μ_{max} features higher join or disjoin dynamics, respectively.

To verify these criteria 4 simulation modes are proposed as illustrated in Table 1. Note that the system parameter N_{max} given in this table represents the *desired group size*, which is needed by some rekeying algorithm to set up the data structures. It differs from the actual supportable n_{max} given in Definition 7.

Table 1. Simulation modes

	Transient	**Scalability**	**Join Dynamics**	**Disjoin Dynamics**
System parameters	$T_J{}^{sys}, T_D{}^{sys}, N_{max}$			
Timing parameters	C_g, C_e, C_h, C_m, C_s			
Group parameters	n_0, λ, μ	λ, μ	n_0, μ	n_0, λ
Simulation parameters	t_{sim}	$T_0, \Delta n,$ $[\, n_{start}, n_{end}\,]$	$T_0, \Delta\lambda,$ $[\, \lambda_{start}, \lambda_{end}\,]$	$T_0, \Delta\mu,$ $[\, \mu_{start}, \mu_{end}\,]$
Variable	time	n	λ	μ
Output metric	$RQoS(t)$ $RAC(t)$	$RQoS_{min}$ RAC_{min}	$RQoS_{min}$ RAC_{min}	$RQoS_{min}$ RAC_{min}

Transient Simulation. This simulation mode estimates the current values of the group size $n(t)$, of the rekeying quality of service $RQoS(t)$, and of the rekeying access control $RAC(t)$. By this means, the behavior of rekeying algorithms over long time periods can be observed. For this purpose, an initial group size n_0, a join rate λ, a disjoin rate μ, and the desired simulation time t_{sim} are set by the user. Similarly to other modes, the transient simulation receives the system parameters T_J^{sys} and T_D^{sys} and the timing parameters (see Definition 22) to estimate the rekeying times RT_J and RT_D. Similarly to the system parameters, the timing parameters are independent of the simulation mode, as can be seen in Table 1. The transient simulation builds the foundation for all other simulation modes.

Scalability Simulation. The importance of this simulation mode results from the significance of the scalability problem in group rekeying. The scalability simulation investigates the effect of the group size on the system behavior, which is implied in the terms RT_J and RT_D of equations (1) and (3). The user sets the group size range $[n_{start}, n_{end}]$ and the simulation step Δn. For each n value of this range, a transient simulation is started with $n_0 = n$. This transient simulation runs over a fixed user-definable observation interval T_o. From all the resulting values of $RQoS$ and RAC for this simulation point, only the worst case values are considered, i.e. $RQoS_{min}$ and RAC_{min}. The scalability simulation helps to estimate the maximal group size n_{max}.

Join/Disjoin Dynamics Simulation. High join rates result in shorter inter-arrival times of join requests and more rekeying computations. This causes longer waiting times for join and disjoin requests according to the terms W_J and W_D in (1) and (3). Thus, higher join rates affect not only $RQoS$, but also RAC. The join dynamic simulation represents a way to investigate theses dependencies. The user defines an initial group size n_0, a disjoin rate μ, and a fixed observation interval T_o. In addition, a simulation range for the join rate λ is entered. For each λ value a transient simulation over T_o is started as in the case of scalability simulation. With the help of join dynamics simulation the maximal join rate λ_{max} for a rekeying algorithm may be determined.

The disjoin dynamics simulation is similar to the join dynamics simulation and can be exploited to calculate the maximal disjoin rate μ_{max}.

4 Rekeying Benchmark Design

The Rekeying Benchmark is mainly composed of two interfaces and three components, as depicted in Figure 3. The *User Interface* (UI) enables users to evaluate different rekeying algorithms by selecting these algorithms and setting the desired parameters. For designers a *Programming Interface* (PI) is provided to integrate new algorithms. In addition, groups with special dynamic behavior, which does not follow Poisson distribution, can be supported by means of a special programming interface module.

The component *Request Generator* creates a rekeying request list depending on the selected group and simulation parameters. An entry of this list keeps information on the request type, join or disjoin, the identity of the member to be

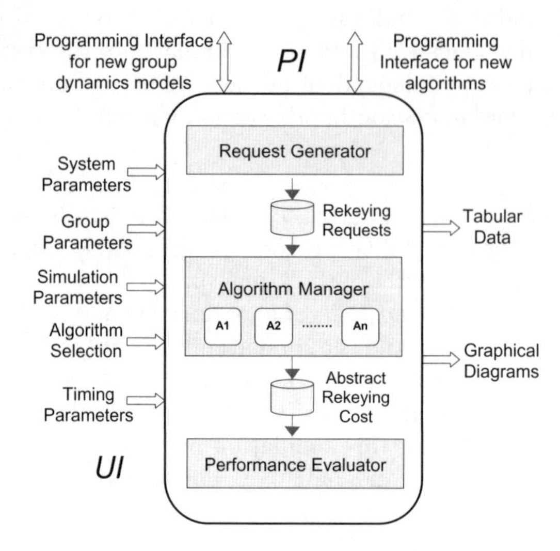

Fig. 3. Benchmark architecture

joined or removed, and the arrival time of this request. The *Algorithm Manager* then selects and configures the rekeying algorithms according to the user settings. It coordinates all the functions of the simulator and controls the rekeying algorithms. Based on the rekeying cost data delivered from the rekeying algorithms and the entered timing parameters, the *Performance Evaluator* finally estimates the rekeying performance of each algorithm and prepares its output for a graphical presentation.

4.1 Request Generator

The Request Generator produces a rekeying request list $RRL(T)$ by executing a main process and 3 subprocesses. Before describing these processes, the following terminology is given.

Definition 9. ***Rekeying Request***
A Rekeying Request is a 3-tuple ($type$, ID, t_a). $type$ indicates the request type, which may be either join (J) or disjoin (D). ID represents the member identity to be joined (ID_J) or removed (ID_D). t_a describes the arrival time of a join (disjoin) request t_{aJ} (t_{aD}), measured from the simulation start point.

Definition 10. ***Rekeying Request List*** $RRL(T)$
A rekeying request list over T, $RRL(T)$, is an ordered set of rekeying requests, which arrive during a defined time interval T. The requests in the list are ordered according to their arrival times.

Example 1. An $RRL(T)$ can be represented in tabular form as seen in Table 2.

Table 2. Example for a rekeying request list $RRL(T)$

Request Type	Member Identity	Arrival Time (ms)
J	1099	0
D	50	0.1
D	178	2
J	22657	5.3

Definition 11. ***Join Arrival List*** $A_J(T)$
A join arrival list over a time interval T is an ordered list of inter-arrival times relating to all join requests generated during T: $A_J(T) = (\Delta t_J(1), \Delta t_J(2), \ldots, \Delta t_J(h))$, where $\Delta t_J(i)$ indicates the inter-arrival time of the i-th join request in the interval T, and

$$\sum_{i=0}^{i=h} \Delta t_J(i) \leq T. \tag{5}$$

Definition 12. ***Disjoin Arrival List*** $A_D(T)$
Similarly to $A_J(T)$, a disjoin arrival list over a time interval T is defined as: $A_D(T) = (\Delta t_D(1), \Delta t_D(2), \ldots, \Delta t_D(k))$, where

$$\sum_{i=0}^{i=k} \Delta t_D(i) \leq T. \tag{6}$$

Definition 13. ***Member Identity*** ID
A member identity is a natural number between 0 and $N_{max} - 1$.

Definition 14. ***Complete Multicast Group*** M
A complete multicast group is the set of all the member identities: $M = ID(i)$, where $i \in 0 \ldots (N_{max} - 1)$.

Definition 15. ***Joined Multicast Subgroup*** M_J
A joined multicast subgroup is the subset of all the given identities. At the start of a simulation with an initial group size n_0, M_J can be given as: $M_J = ID(i)$, where $i \in 0 \ldots (n_0 - 1)$.

Definition 16. ***Potential Multicast Subgroup*** M_D
A potential multicast subgroup is the subset of all the identities, which can be given to new members. At the start of a simulation with an initial group size n_0, M_D can be given as: $M_D = ID(i)$, where $i \in n_0 \ldots (N_{max} - 1)$.

Algorithm 1. GenReqList

Require: T
Ensure: $RRL(T)$

1: GetArrivalLists(T)$\rightarrow A_J(T)$ and $A_D(T)$
2: $i := 1$, $j := 1$, $t_{aJ} := 0$, $t_{aD} := 0$
3: **while** $i \leq h$ or $j \leq k$ **do**
4:   **if** $\Delta t_J(i) \geq \Delta t_D(j)$ **then**
5:     $t_{aD} := t_{aD} + \Delta t_D(j)$
6:     GetDisjoinID $\rightarrow ID_D$
7:     Add (D, ID_D, t_{aD}) into $RRL(T)$
8:     $j := j + 1$
9:   **else**
10:     $t_{aJ} := t_{aJ} + \Delta t_J(i)$
11:     GetJoinID $\rightarrow ID_J$
12:     Add (J, ID_J, t_{aJ}) into $RRL(T)$
13:     $i := i + 1$
14:   **end if**
15: **end while**
16: Sort $RRL(T)$ according to increasing arrival times.
17: **return** $RRL(T)$

Request Generator Process GenReqList. This process generates the rekeying request list $RRL(T)$ as given in Algorithm 1. First, the arrival process GetArrivalLists(T) is called to produce join and disjoin arrival lists $A_J(T)$ and $A_D(T)$. According to the inter-arrival times in these lists, the arrival times for the individual requests are determined. Depending on the request type, the member identity is then obtained by calling GetJoinID or GetDisjoinID. Then, the $RRL(T)$ is updated by the new request. After processing all entries in $A_J(T)$ and $A_D(T)$, the $RRL(T)$ is sorted with increasing arrival time. Note that the request generator is transparent to the simulation mode. Utilizing the generator for different simulation modes will be described in the scope of the Algorithm Manager. Example 2 illustrates this code in more details.

Example 2. Assume a group of maximal 8 members, where 5 members are currently joined as follows: $M = \{0, 1, 2, 3, 4, 5, 6, 7\}$, $M_J = \{0, 1, 2, 3, 4\}$, $M_D = \{5, 6, 7\}$, See definitions 14, 15, and 16 for M, M_J and M_D, respectively. Assume that calling the process GetArrivalLists(T) on some interval T results in the inter-arrival time lists $A_J(T) = (10, 25)$ and, $A_D(T) = (11, 5, 7)$. This means that during the given interval 2 join requests and 3 disjoin requests are collected, i.e. $h = 2$, $k = 3$. In addition, the requests feature the following inter-arrival times: $\Delta t_J(1) = 10$, $\Delta t_J(2) = 25$, $\Delta t_D(1) = 11$, $\Delta t_D(2) = 5$, and $\Delta t_D(3) = 7$.

In the first run of the while loop, the if-condition in Algorithm 1 is false because $\Delta t_J(1) < \Delta t_D(1)$. Therefore, the first join request is processed by determining its arrival time as $t_{aJ} := 0 + 10 = 10$, as $t_{aJ} = 0$ initially. Assuming that executing the process GetJoinID returned a member identity $ID_J = 5$, a first entry is written into the rekeying request list $RRL(T)$, as depicted in the first row of Table 3 which represents the $RRL(T)$ for this example. In the second iteration the if-condition is true because $\Delta t_J(2) > \Delta t_D(1)$. Therefore, the next request to be written to the $RRL(T)$ is of a disjoin type and has an arrival time $t_{aD} := 0 + 11 = 11$, as $t_{aD} = 0$ initially. Assuming that GetDisjoinID returns an ID_D which is equal to 3, the $RRL(T)$ is extended by the second entry of Table 3. The other entries of Table 3 can be determined in the same way. Figure 4 illustrates the relation between the inter-arrival times generated by the process GetArrivalList(T) and the estimated arrival times in the given example.

Table 3. $RRL(T)$ of Example 2

Request Type	Member Identity	Arrival Time (ms)
J	5	10
D	3	11
D	1	16
D	4	23
J	1	35

Arrival Process GetArrivalLists. Based on related work on modeling multicast member dynamics [15], the rekeying simulator assumes inter-arrival times, which follow an exponential distribution for join and disjoin requests with the rates λ and μ respectively. The corresponding cumulative distribution functions are given by:

$$F_J(\Delta t_J) = 1 - e^{-\lambda \Delta t_J} \qquad F_D(\Delta t_D) = 1 - e^{-\mu \Delta t_D} \tag{7}$$

To generate an exponentially distributed random variate based on uniform random numbers between 0 and 1, the inverse transformation technique can be used. Accordingly, if r represents such a random number, the inter-arrival times of a join and disjoin request can be estimated as:

$$\Delta t_J = -\frac{1}{\lambda} \ln r \qquad \Delta t_D = -\frac{1}{\mu} \ln r \tag{8}$$

Algorithm 2 outlines the arrival process for creating the join arrival list $A_J(T)$. Creating $A_D(T)$ is identical and omitted, for brevity.

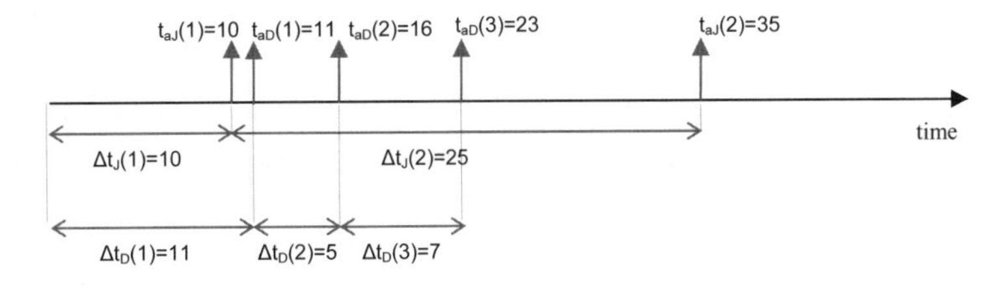

Fig. 4. Arrival times and inter-arrival times for Example 2

Algorithm 2. GetArrivalLists

Require: T
Ensure: $A_J(T)$ (Generating $A_D(T)$ is identical)
1: $\sum \Delta t_J := 0; \sum \Delta t_D := 0;$
2: **while** $\sum \Delta t_J \leq T$ **do**
3: Generate r
4: Determine $\Delta t_J = 0$ according to (8)
5: $\sum \Delta t_J = \sum \Delta t_J + \Delta t_J$
6: Add Δt_J to $A_J(T)$
7: **end while**
8: **return** $A_J(T)$

Join/Disjoin Identity Selection Processes GetJoinID and GetDisjoinID. To join a member, any identity ID_J may be selected from the potential multicast subgroup M_D. A possible selection strategy may rely on choosing the smallest available ID_J, which allows some order in the group management. In contrast, selecting a leaving ID_D from M_J is inherently non-deterministic, as a group owner can not forecast which member will leave the group. To select an ID_D the following method is proposed. The ID_D's of M_J are associated with continuous successive indices from 0 to $m-1$, where m is the number of all ID_D's in M_J. To select an ID_D, first a uniform zero-one random number r is generated. Then an index i is determined as $m \cdot r$. In a last step, the ID_D is selected, which has the index i.

4.2 Algorithm Manager

The algorithm manager plays a central role in the benchmark architecture. Its functionality can be illustrated by the process described in Figure 5. After reading the user settings of the desired parameters, the simulation mode, and the algorithms to be evaluated, the algorithm manager executes the corresponding simulation process. Simulation processes on their part call the request generator and pass the rekeying requests to the selected rekeying algorithms. As a result, a simulation process provides abstract rekeying costs, i.e. without timing information. This information is first supplied to the performance evaluator which determines the metrics $RQoS$ and RAC. In this section the underlying simulation processes DoTranSim, DoScalSim, DoJoinDynSim, and DoDisjoinDynSim will be explained. For this purpose, three basic concepts are introduced first.

Definition 17. ***Abstract Rekeying Cost (ARC)***Abstract rekeying cost is a 5-tuple (G, E, H, M, S), which specifies a rekeying request and gives the numbers of cryptographic operations needed to grant this request by a rekeying algorithm. The elements of the ARC are specified in Table 4.

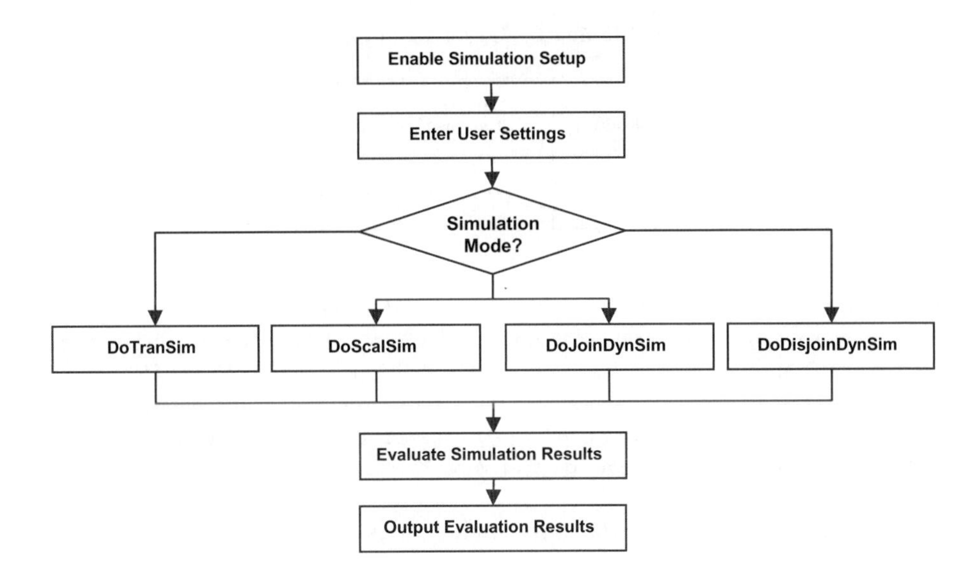

Fig. 5. Rekeying performance evaluation procedure

Table 4. Abstract rekeying costs notation

ARC Element	Meaning
G	# Generated cryptograpic keys
E	# Symmetric encryptions
H	# Cryptographic hash operations
M	# Message authentication codes
S	# Digital signatures

Definition 18. ***Rekeying Cost List RCL(T)***
A rekeying cost list is a rekeying request list $RRL(T)$, see Definition 10, which is extended by the abstract rekeying cost ARC for each request. $RCL(T)$ is used for transient simulation.

Example 3. Table 5 shows an example for an $RCL(T)$, which is an extension of the rekeying request list given in Table 2. This example results from executing the LKH algorithm with binary trees. This can be seen from the fact that each generated key is encrypted twice to determine the rekeying submessages. Note that the rekeying algorithm in this example does not use group authentication. Therefore, no message authentication codes are needed. Instead, rekeying submessages are hashed and the final hash value is signed once for each request [18].

Table 5. RCL(T) for Example 3

Request Type	Member Identity	Arrival Time (ms)	Rekeying Cost List RCL(T)				
			G	E	H	M	S
L	1099	0	6	12	12	0	1
J	50	0.1	3	6	6	0	1
J	178	2	8	16	16	0	1
L	22657	5.3	2	4	4	0	1

Definition 19. ***Complex Rekeying Cost List CRCL(T)***
A complex rekeying cost list over an interval T is a set of rekeying cost lists generated over this interval under different group conditions: $CRCL(T) = \{RCL1(T), RCL2(T), \cdots\}$. $CRCL(T)$ is used in the simulation modes of scalability and join/disjoin dynamics, where an $RCL(T)$ is generated for each n, λ, or μ value in the desired simulation range, respectively.

Transient Simulation. Algorithm 3 represents the process of transient simulation DoTranSim. The request generator process is resumed to generate a request list $RRL(t_{sim})$ for the desired simulation time period. For each selected rekeying algorithm, the Algorithm Manger performs two main steps. Firstly, the rekeying algorithm is requested to initialize a group with n_0 members. Secondly, each request of $RRL(t_{sim})$ is sent to the rekeying algorithm, which determines the corresponding abstract rekeying cost ARC for that request.

Algorithm 3. DoTranSim

Require: All settings for a transient simulation as given in Table 1; set of rekeying algorithms to be evaluated.
Ensure: A $RCL(t_{sim})$ for each rekeying algorithm
1: GenReqList(t_{sim}) according to Algorithm 1 $\rightarrow RRL(t_{sim})$
2: **for** each algorithm **do**
3: Initialize the group with n_0 members
4: **while** $RRL(t_{sim})$ is not empty **do**
5: Send a rekeying request to the algorithm
6: Get corresponding ARC
7: Add ARC to $RCL(t_{sim})$
8: **end while**
9: **end for**
10: **return** $RCL(t_{sim})$ for all algorithms

Other Simulation Modes. As mentioned in Section 3.2, other simulation modes are highly similar and rely all on the transient simulation mode. Therefore, only the scalability simulation is given in Algorithm 4, for brevity.

Algorithm 4. DoScalSim

Require: All settings for a scalability simulation as given in Table 1; set of rekeying algorithms to be evaluated.
Ensure: A $CRCL(T_o)$ for each rekeying algorithm
1: **for** each algorithm **do**
2: $n := n_{start}$
3: **while** $n \leq n_{end}$ **do**
4: DoTranSim for T_o and $n_0 = n$ according to Algorithm 3$\rightarrow RCL(T_o)$
5: Add $RCL(T_o)$ to $CRCL(T_o)$
6: $n := n + \Delta n$
7: **end while**
8: **end for**
9: **return** $CRCL(T_o)$ for all algorithms

4.3 Performance Evaluator

This component receives a set of $RCL(T)$ or $CRCL(T)$ and calculates the system metrics $RQoS$ and RAC as a function of time, group size, join rate, or disjoin rate.

Definition 20. ***Performance Simulation Point (PSP)***
A performance simulation point is a 3-tuple (x, $RQoS$, RAC), where x is the variable, which the $RQoS$ and RAC are related to, e.g. n in the scalability simulation. Depending on the simulation mode x, $RQoS$, and RAC are interpreted as illustrated in Table 1. Note that in a transient simulation $RQoS$ and RAC are not defined for a disjoin and join request, respectively.

Definition 21. ***Rekeying Performance List (RPL)***
A rekeying performance list is a set of performance simulation points. $RPL = \{PSP\} = \{(x_1, RQoS_1, RAC_1), (x_2, RQoS_2, RAC_2), \ldots\}$.

Definition 22. ***Timing Parameter List (TPL)***
A timing parameter list (TPL) is a 5-tuple (C_g, C_e, C_h, C_m, C_s), where the tuple elements are defined as given in Table 6. Recall that the timing parameters reflect the performance of cryptographic algorithms and of the platform on an abstraction level, which allows for a reliable evaluation of different rekeying algorithms.

Table 6. Timing parameters

TPL Element	Meaning: Time cost of
C_g	Generating one cryptograpic key
C_e	One symmetric encryption
C_h	One cryptographic hash operation
C_m	One message authentication code
C_s	One digital signature

The Performance Evaluator executes processes, which combine a rekeying cost list $RCL(T)$ or a complex rekeying cost list $CRCL(T)$ with a timing parameter list TPL to produce a rekeying performance list PRL for a specific rekeying algorithm. For each rekeying request in $RCL(T)/CRCL(T)$ the actual join/disjoin time is established according to equations (1) and (3). The rekeying and waiting times for a join or disjoin request are determined as

$$RT_{J/D} = G \cdot C_g + E \cdot C_e + H \cdot C_h + M \cdot C_m + S \cdot C_s \tag{9}$$

$$W_{J/D} = \begin{cases} \sum_{i=1}^{m} RT_i & \text{if } m \geq 1 \\ 0 & \text{if } m = 0 \end{cases} \tag{10}$$

where m represents the number of all requests waiting in the system queue or being processed at the arrival of the request at hand. Knowing the waiting times and the rekeying times, the actual rekeying times can be estimated using (1) and (3). Afterwards, $RQoS$ and RAC can be calculated for a join or disjoin request according to (2) or (4), respectively.

Transient Evaluation Process (EvalTranSimResults). In the case of a transient simulation the performance evaluator executes the process EvalTranSimResults according to Algorithm 5. For each join and disjoin request in the $RCL(T)$, a performance simulation point PSP is determined. The symbol ∞ in the pseudo code indicates an undefined metric for the current request. For example, $RQoS$ is not defined for a disjoin request. t_{aJ} and t_{aD} represent the arrival times of the corresponding join and disjoin requests, respectively. Remember that these time values are determined from the arrival lists by the request generator process according to Algorithm 1.

Algorithm 5. EvalTranSimResults

Require: A $RCL(t_{sim})$ for each rekeying algorithm, T_J^{sys}, T_D^{sys}
Ensure: A PRL for each rekeying algorithm
1: **for** each $RCL(t_{sim})$ **do**
2:   **for** each request in $RCL(t_{sim})$ **do**
3:     **if** request type = J **then**
4:       Determine RT_J and W_J according to (9) and (10)
5:       Determine T_J and $RQoS$ according to (1) and (2)
6:       PSP=(t_{aJ}, $RQoS$, ∞)
7:     **else**
8:       Determine RT_D and W_D according to (9) and (10)
9:       Determine T_D and RAC according to (3) and (4)
10:       PSP=(t_{aD}, ∞, RAC)
11:     **end if**
12:     Add PSP to PRL
13:   **end for**
14: **end for**
15: **return** PRL for all algorithms

Complex Evaluation Process (EvalComplexSimResults). Other simulation modes deliver a $CRCL(T)$. The Performance Evaluator generates one performance simulation point PSP for each $RCL(T)$ of $CRCL(T)$. The first element of the PSP tuple represents a n, λ, or μ value for scalability, join dynamics or disjoin dynamics simulation, respectively. The second element represents the minimal rekeying quality of service $RQoS_{min}$ of all join requests in the observation time for the corresponding n, λ, or μ value. Similarly, the third element represents RAC_{min} of all disjoin requests. Algorithm 6 depicts the process EvalComplexSimResults for evaluating non-transient simulation results. The symbol ∞ in this pseudo code indicates an initial very large value of the corresponding metric.

Algorithm 6. EvalComplexSimResults

Require: A $CRCL(T_o)$ for each rekeying algorithm, T_J^{sys}, T_D^{sys}
Ensure: A PRL for each rekeying algorithm
1: **for** each rekeying algorithm **do**
2:   **for** each $RCL(T_o)$ of $CRCL(T_o)$ **do**
3:     $RQoS_{min} = \infty$, $RAC_{min} = \infty$
4:     **for** each request in $RCL(T_o)$ **do**
5:       **if** request type = J **then**
6:         Determine RT_J and W_J according to (9) and (10)
7:         Determine T_J and $RQoS$ according to (1) and (2)
8:         **if** $RQoS < RQoS_{min}$ **then**
9:           $RQoS_{min} := RQoS$
10:         **end if**
11:       **else**
12:         Determine RT_D and W_D according to (9) and (10)
13:         Determine T_D and RAC according to (3) and (4)
14:         **if** $RAC < RAC_{min}$ **then**
15:           $RAC_{min} := RAC$
16:         **end if**
17:       **end if**
18:       PSP=($n/\lambda/\mu$, $RQoS_{min}$, RAC_{min})
19:     **end for**
20:     Add PSP to PRL
21:   **end for**
22: **end for**
23: **return** PRL for all algorithms

5 Case Study

LKH is a tree-based rekeying scheme. As an effect of multiple disjoin processes, the key tree may get out of balance. Several solutions have been proposed to rebalance the tree in this case. The first contribution originates from Moyer [11] who introduced two methods to rebalance the key tree, an immediate and a periodic rebalancing. Only a cost analysis after one disjoin request is given for the first method. The periodic rebalancing is not analyzed. In [12] a method for rebalancing based on sub-trees was presented. A comparison with the solution of Moyer is drawn, but not with the original LKH. In [13] an AVL-tree rebalancing methods was applied to key trees. However, no backward access control is guaranteed in this solution. In [14] three algorithms for tree rebalancing were proposed. Simulation results are provided, which assume equally likely join and disjoin behavior. However, this condition itself ensures tree balancing, because a new member can be joined at the leaf of the most recently removed member. The same applies to the simulation results by Lu [9]. From this description, it is obvious that a comprehensive analysis is needed to justify the employment of rebalancing, which is associated with extra rekeying costs resulting from shifting members between tree leaves. The rekeying benchmark offers this possibility by allowing a simultaneous evaluation of two LKH algorithms (with and without rebalancing) under complex conditions. Especially the effect of the disjoin rate is of interest in case of rebalancing. Therefore, a disjoin dynamics simulation is performed under the following conditions: $T_J^{sys} = T_D^{sys} = 100ms$, $N_{max} = 65.536$, $C_g = C_e = C_h = C_m = 1\mu s$, $C_s = 15ms$, $n_0 = 4096$, $T_o = 1s$, $\lambda = 10s^{-1}$, $\mu_{start} = 1s^{-1}$, $\mu_{stop} = 20s^{-1}$, $\Delta\mu_{start} = 1s^{-1}$. Simulation results are depicted in Figure 6 and Figure 7.

These diagrams clearly unveil that rebalancing degrades both *RQoS* and *RAC* values and that this degrading increases with an increasing disjoin rate. Thus, the simulation discloses that additional rekeying costs associated with rebalancing exceed the performance gain achieved by it. Consequently, rebalancing is not advantageous for LKH trees, at least under the given simulation conditions.

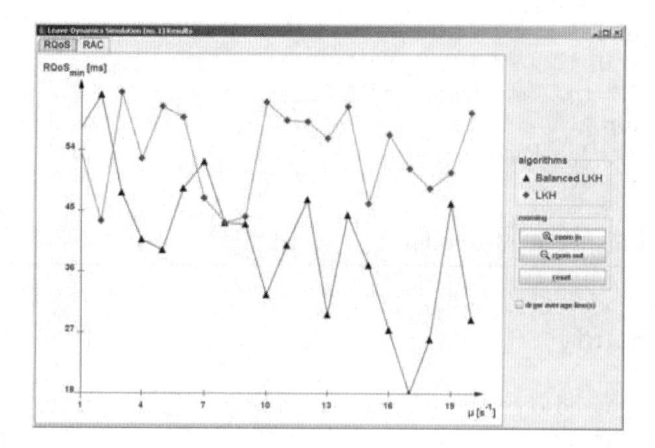

Fig. 6. RQoS in rebalanced vs. non-rebalanced LKH

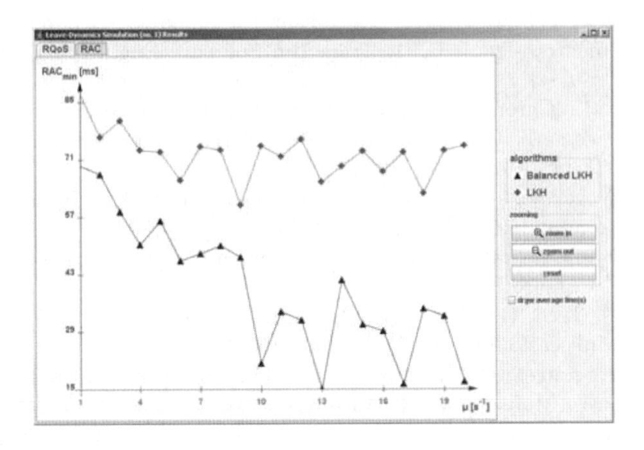

Fig. 7. RAC in rebalanced vs. non-rebalanced LKH

6 Conclusion

An assessment methodology and an associated simulation tool were presented as a novel method to deal with the rekeying performance evaluation problem. By means of the underlying concept of abstraction a reliable and meaningful evaluation of different rekeying algorithms is provided. A case study illustrated the advantage of this benchmark in analyzing yet unanswered questions relating to rekeying. In its first prototype, the benchmark considers rekeying costs in terms of cryptographic operations to be run on the server side. Other cost factors, such as tree traversing in LKH, will be addressed in future work. Additionally, more rekeying algorithms will be programmed and evaluated.

References

1. Wong, C.K., Gouda, M., Lam, S.S.: Secure Group Communication Using Key Graph. IEEE/ACM Trans. on Networking 8(1), 16–30 (2000)
2. Ng, W.H.D., Sun, Z.: Multi-Layers Balanced LKH. In: Proc. of IEEE Int. Conf. on Communication ICC, pp. 1015–1019 (May 2005)
3. Li, X.S., Yang, Y.R., Gouda, M., Lam, S.S.: Batch Rekeying for Secure Group Communications. In: Proc. ACM 10th Int. World Wide Web Conf., Hong Kong (May 2001)
4. Amir, Y., Kim, Y., Nita-Rotaru, C., Tsudik, G.: On the Performance of Group Key Agreement Protocols. ACM Trans. on Information Systems Security 7(3), 457–488 (2004)
5. Pegueroles, J., Rico-Novella, F.: Balanced Batch LKH: New Proposal, Implementation and Performance Evaluation. In: Proc. IEEE Symp. on Computers and Communications, p. 815 (2003)
6. Chen, W., Dondeti, L.R.: Performance Comparison of Stateful and Stateless Group Rekeying Algorithms. In: Proc. of Int. Workshop in Networked Group Communication (2002)

7. Sherman, A., McGrew, D.: Key Establishment in Large Dynamic Groups Using One-Way Function Trees. IEEE Trans. on Software Engineering 29(5), 444–458 (2003)
8. Waldvogel, M., Caronni, G., Sun, D., Weiler, N., Plattner, B.: The VersaKey Framework: Versatile Group Key Management. IEEE J. on Selected Areas in Communications 17(8), 1614–1631 (1999)
9. Lu, H.: A Novel High-Order Tree for Secure Multicast Key Management. IEEE Trans. on Computers 54(2), 214–224 (2005)
10. Mittra, S.: Iolus: A Framework for Scalable Secure Multicasting. In: Proc. of ACM SIGCOMM, Cannes, France, pp. 277–288 (September 1997)
11. Moyer, M.J., Tech, G., Rao, J.R., Rohatgi, P.: Maintaining Balanced Key Trees for Secure Multicast, Internet draft (June 1999), `http://www.securemulticast.org/draft-irtf-smug-key-tree-balance-00.txt`
12. Moharrum, M., Mukkamala, R., Eltoweissy, M.: Efficient Secure Multicast with Well-Populated Multicast Key Trees. In: Proc. of IEEE ICPADS, p. 214 (July 2004)
13. Rodeh, O., Birman, K.P., Dolev, D.: Using AVL Trees for Fault Tolerant Group Key Management, Tech. Rep. 2000-1823, Cornell University (2000)
14. Goshi, J., Ladner, R.E.: Algorithms for Dynamic Multicast Key Distribution Trees. In: Proc. of ACM Symp. on Principles of Distributed Computing, pp. 243–251 (2003)
15. Almeroth, K.C., Ammar, M.H.: Collecting and Modeling the join/leave Behavior of Multicast Group Members in the MBone. In: Proc. of HPDC, pp. 209–216 (1996)
16. NIST (National Institute of Standards and Technology), Advanced Encryption Standard (AES), Federal Information Processing Standard 197 (November 2001)
17. `http://www.cryptopp.com/`
18. Shoufan, A., Laue, R., Huss, S.A.: High-Flexibility Rekeying Processor for Key Management in Secure Multicast. In: IEEE Int. Symposium on Embedded Computing SEC 2007, Niagara Falls, Canada (May 2007)

Robustness Analysis of Watermark Verification Techniques for FPGA Netlist Cores

Daniel Ziener, Moritz Schmid, and Jürgen Teich

Hardware/Software Co-Design
Department of Computer Science
University of Erlangen-Nuremberg, Germany
Am Weichselgarten 3
91058 Erlangen, Germany
{daniel.ziener,moritz.schmid,teich}@cs.fau.de

Abstract. In this paper we analyze the robustness of watermarking techniques for FPGA IP cores against attacks. Unlike most existing watermarking techniques, the focus of our techniques lies on ease of verification, even if the protected cores are embedded into a product. Moreover, we have concentrated on higher abstraction levels for embedding the watermark, particularly at the logic level, where IP cores are distributed as netlist cores. With the presented watermarking methods, it is possible to watermark IP cores at the logic level and identify them with a high likelihood and in a reproducible way in a purchased product from a company that is suspected to have committed IP fraud. For robustness analysis we enhanced a theoretical watermarking model, originally introduced for multimedia watermarking. Finally, two exemplary watermarking techniques for netlist cores using different verification strategies are described and the robustness against attacks is analyzed.

1 Introduction

The ongoing miniaturization of on-chip structures allows us to implement very complex designs which require very careful engineering and an enormous effort for debugging and verification. Indeed, complexity has risen to such enormous measures that it is no longer possible to keep up with productivity constraints if all parts of a design must be developed from scratch. In addition, the very lively market for embedded systems with its demand for very short product cycles intensifies this problem significantly. A popular solution to close this so called productivity gap is to reuse design components that are available in-house or that have been acquired from other companies. The constantly growing demand for ready to use design components, also known as IP cores, has created a very lucrative and flourishing market which will continue its current path not only into the near future.

One problem of IP cores is the lack of protection mechanisms against unlicensed usage. A possible solution is to hide a unique signature (watermark) inside the core by which the original author can be identified and an unlicensed

A. Biedermann and H. Gregor Molter (Eds.): Secure Embedded Systems, LNEE 78, pp. 105–127.
springerlink.com

usage can be proven. Our vision is that it should be possible to detect the unlicensed usage of an IP core solely using the product in which the IP core may be embedded and the watermark information of the original author. It should not be necessary to request any additional information from the manufacturer of suspicious product. Such concepts of course need advanced verification techniques in order for a signature or certain characteristics to be detectable in one of possibly many IP cores inside a system. Another aspect to be considered is the fact that IP cores will undergo several sophisticated optimization steps during the course of synthesis. It is of utmost importance that a watermark is transparent towards design and synthesis tools, that is, the embedded identification must be preserved in all possible scenarios. Whilst on the one hand, we must deal with the problem that automated design tools might remove an embedded signature all by themselves, a totally different aspect is that embedded signatures must also be protected against the removal by illegitimate parties whose intention is to keep the IP core from being identifiable. The latter is not to be taken lightly because if a sufficiently funded company decides to use unlicensed cores to, for example, lower design costs, there are usually very high skilled employees assigned with the task to remove or bypass the embedded watermark.

Figure 1 depicts a possible watermarking flow. An IP core developer embeds a signature inside his IP core using a watermark embedder and publishes the protected IP core. The intention of this procedure is that companies interested into using the developer's core would obtain a licensed copy. However, a third-party company may also obtain an unlicensed copy of the protected IP core and use it in one of their products. If the IP core developer becomes suspicious that his core might have been used in a certain product without proper licensing, he can simply acquire the product and check for the presence of his signature. If this attempt is successful and his signature presents a strong enough proof of authorship, the developer may decide to accuse the product manufacturer of IP fraud and press legal charges.

IP cores exist for all design flow levels, from plain text HDL cores on the register-transfer level (RTL) to bitfile cores for FPGAs or layout cores for ASIC designs on the device level. In the future, IP core companies will concentrate more and more on the versatile HDL and netlist cores due to their flexibility. One reason for this development is that these cores can be easily adapted to new technologies and different FPGA devices. This work focuses on watermarking methods for IP cores implemented for FPGAs. These have a huge market segment and the inhibition threshold for using unlicensed cores is lower than in the ASIC market where products are produced in high volumes and vast amounts of funds are spent for mask production. Moreover, we concentrate on flexible IP cores which are delivered on the logic level in a netlist format. The advantage of this form of distribution is that these cores can be used for different families FPGA devices and can be combined with other cores to obtain a complete SoC solution. Our work differs from most other existing watermarking techniques, which do not cover the area of HDL and netlist cores, or are not able to easily

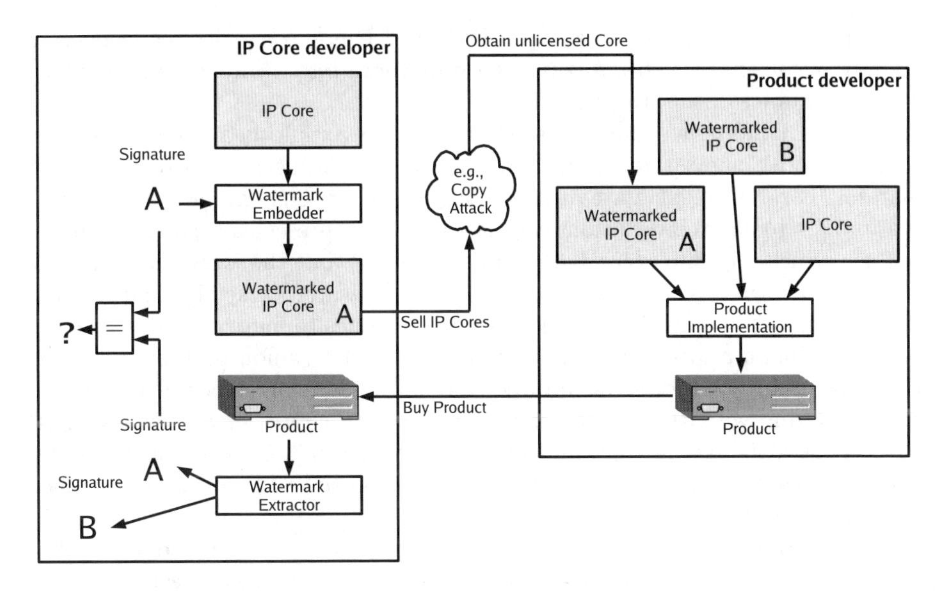

Fig. 1. An IP core developer embeds a watermark inside his core. If a company uses this core in one of their product without proper licensing, the IP core developer can obtain the product and check it for the presence of his watermark.

extract an embedded watermark from a heterogeneous SoC implemented in a given product.

The remaining work is organized as follows: In Section 2, a short overview of related work for IP watermarking is provided. Afterwards, Section 3 presents a theoretical model for watermarking IP cores. Section 4 deals with different strategies to extract a watermark from an FPGA embedded into a product. We proceed by describing two methods for extracting a watermark. The first method explains the extraction of a watermark from an FPGA bitfile in Section 5. Analyzing the power consumption of the FPGA in order to verify the presence of a watermark is the second method and will be discussed in Section 6. Additionally, the robustness against typical attacks will be analyzed for both methods. In conclusion, the contributions will be summarized.

2 Related Work

IP cores are often distributed like software and can therefore be used without proper legitimacy, which ranges from over-provisioning agreed on amounts of licensed uses to simply not licensing an IP core at all. Some core suppliers use encryption to protect their cores. The downside of such approaches is that encrypted cores can only be used in conjunction with special tools and that the encryption will eventually be broken by individuals with high criminal energy. A different approach is to hide a signature in the core, a so-called watermark, which can be used as a reactive proof of the original ownership enabling IP core developers to identify and react upon IP fraud. There exist many concepts and approaches on the issue of integrating a watermark into a core, several of which will be reviewed in this section.

In general, hiding a signature into data, such as a multimedia file, some text, program code, or even an IP core by steganographic methods is called watermarking. For multimedia data, it is possible to exploit the imperfection of human eyes or ears to enforce variations on the data that represent a certain signature, but for which the difference between the original and the watermarked work cannot be recognized. Images, for example, can be watermarked by changing the least significant bit positions of the pixel tonal values to match the bit sequence of the original authors signature. For music, it is a common practice to watermark the data by altering certain frequencies, the ear cannot perceive and thus not interfering with the quality of the work [5]. In contrast, watermarking IP cores is entirely different from multimedia watermarking, because the user data, which represents the circuit, must not be altered since functional correctness must be preserved.

Most methods for watermarking IP cores focus on either introducing additional constraints on certain parts of the solution space of synthesis and optimization algorithms, or adding redundancies to the design.

Additive methods add a signature to the functional core, for example, by using empty lookup-tables in an FPGA [15,19] or by sending the signature as a preamble of the output of the test mode [8]. Constraint-based methods were originally introduced by [9] and restrict the solution space of an optimization algorithm by setting additional constraints which are used to encode the signature. Methods for constraint-based watermarking in FPGAs exploit the scan-chain [13], preserve nets during logic synthesis [12], place constraints for CLBs in odd/even rows [10], alter the transistor width [4] or route constraints with unusual routing resources [10].

A common problem of many watermarking approaches is that for verification of the presence of the marks, the existence and the characteristic of a watermark must be disclosed, which enables possible attackers to remove the watermark. To overcome this obstacle, Adelsbach [2] and Li [16] have presented so-called zero-knowledge watermark schemes which enable the detection of the watermark without disclosing relevant information.

A survey and analysis of watermarking techniques in the context of IP cores is provided by Abdel-Hamid and others [1]. Further, we refer to our own survey of watermarking techniques for FPGA designs [25]. Moreover, a general survey of security topics for FPGAs is given by Drimer [7].

3 Theoretical Watermark Model for Robustness Analysis against Attacks

In this section, we propose a theoretical model for IP core watermarking. With this model, different threats and attack scenarios can be described and evaluated. In general, watermarking techniques must deal with an uncontrolled area, where the watermarked work is further processed. This is true for multimedia watermarking, where, for example, watermarked images are processed to enhance the image quality by filters or for IP core watermarking where the core is combined with other cores and traverses other design flow steps. However, the watermarked work may be exposed to further attacks in this uncontrolled area that may destroy the watermark and thus the proof of authorship as well. This uncontrolled area is difficult to describe in a precise way and therefore, the security goals and issues for watermarking are often given in natural language which results in an imprecise description. This natural description makes an assessment of the security very difficult, particularly if the attackers are intelligent and creative.

Introducing a defined theoretical watermarking model with attackers and threats allows us to assess the security of IP core watermarking techniques. However, it should be noted that the model has to cover all possible attack scenarios and represent all aspects of the real world behavior to allow for a meaningful assessment of the security. In this section, we present a general watermark model introduced by Li et al. [17] which will be enhanced with aspects of IP core watermarking.

Watermarking intellectual property can be specified precisely by characterizing the involved actions using a security model. We use the standard definitions from security theory, which defines security goals, threats and attacks. Security goals represent certain abilities of a scheme, which are important to protect in order to keep its functionality in tact. These abilities may be violated by threats which are realized by attacks. Regarding watermarking, the overall security goal is to be able to present a proof of authorship that is strong enough to hold in front of a court. The security goal of a watermark scheme is violated if the original author cannot produce a strong enough proof of authorship, so that a dispute with another party will lead to an ownership deadlock, but also in the occasion, where another party is able to present a more convincing proof of authorship than the original author, resulting in counterfeit ownership. Another violation of the proof of authorship occurs if the watermark of a credible author is forged by another author and is used to convince a third party, that a work was created by someone who did not.

An attacker can realize an ownership deadlock, if he can present a watermark in the work, that is at least as convincing as the original authors watermark. If such an ambiguity attack is successful, the real ownership cannot be decided and the original author cannot prove his authorship. If, in addition, the ambiguity attack results in the pirate being able to present an even more convincing proof of authorship than the creator of the work, the pirate can counterfeit the ownership. Another way to take over the ownership of a piece of IP is to be able to remove the original authors watermark by means of a removal attack. Forged authorship can be achieved by a key copy attack which simply duplicates the means of creating a credible authors watermark. One last violation of the security goal does not directly involve the author, but requires him to not take part in a dispute over theft. The theft of a work resulting in counterfeit ownership can be realized by

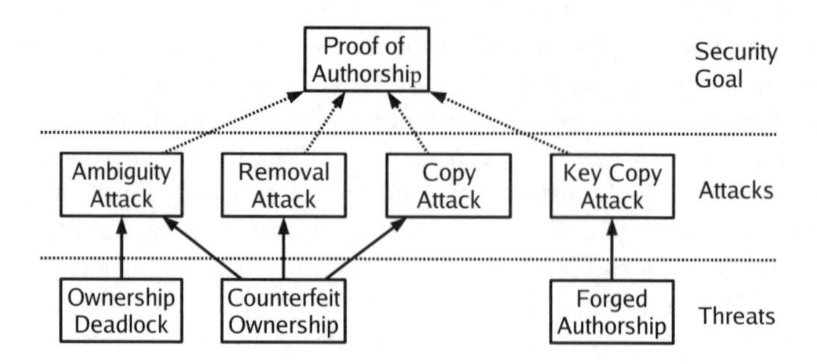

Fig. 2. An overview of threats, attacks and the watermarking security goal of the proof of authorship. The different threats are realized by attacks which violate the security goal.

a copy attack. The realized threat is only successful until the original author realizes the violation. An overview of the introduced terms can be observed in Figure 2.

Watermarking IP cores in electronic design automation is in some aspects different from multimedia watermarking (image, audio, etc.). An essential difference is that watermarking must preserve the functionality of the core. Another difference is that IP cores can be distributed at several abstraction levels which have completely different properties for the watermark security against attacks. We define different design steps as different technology or abstraction levels a work or IP core can be specified on. On higher abstraction levels, as for example on the architecture or register-transfer level, the functionality is described by an algorithm. At these levels, mainly the behavior is described and the representation is optimized for easy reading and understanding the algorithm. During the course of the design flow, more and more information is added. For example, placement information is included at the device level representation of the core. Extracting only the relevant information about the behavior of the algorithm is much harder than at higher abstraction levels. Furthermore, the information at lower abstraction levels is usually interpreted by tools rather than humans. The representation of this information is therefore optimized for machine and not human readability. For example, consider an FPGA design flow. Here, three different abstraction levels exist: RTL, logic, and device level. An algorithm, specified on the register-transfer-level (RTL) in an HDL core is easier to understand than a synthesized algorithm on the logic level, represented by a netlist. In summary, we can say that the behavior of an algorithm is easier to understand on higher abstraction levels than it is on the lower ones.

Transformations from a higher to a lower abstraction level are usually done by design tools. For example, a synthesis tool is able to transform an HDL core specified on the register-transfer level (RTL) into its representation on the logic level. Transformations from a lower to a higher level can be achieved by reverse engineering. Here, usually no common tools are available. One exception is the Java library *JBits* from Xilinx [21] which is able to interpret the bitfiles of Virtex-II device types. Thus, it is possible to transfer a bitfile core into a netlist at the logic level by using JBits. However, in general, reverse engineering must be considered as very challenging task which may cause high costs.

A watermark can be embedded at every abstraction level. Furthermore, the watermarked core can be published and distributed also at every abstraction level which must not necessarily be the same level at which the watermark was embedded. However, the extraction of the watermark is usually done on the lowest abstraction level, because this is the representation of the design which is implemented into the end product.

Hiding a watermark at a lower abstraction level is easier, because first, there are more possibilities of how and where to hide the watermark and second, the information stored at these abstraction levels is usually outside the reception area of the human developer.

To explain all threats and attacks in detail, some definitions have to be made first [17,20].

3.1 Definitions

Our definitions for the IP core watermarking model [20] are derived from the general watermarking model introduced by Li et al. [17]. A work or IP core that is specified at abstraction level Y is denoted by $I_Y = (x_{Y_1}, x_{Y_2}, \ldots, x_{Y_m})$, where each $x_{Y_i} \in \mathcal{I}_Y$ is an element of the work, and $\mathcal{I}_Y$ is a universe, inherent to the abstraction level Y. For example, an FPGA design at the device abstraction level might be represented by a bitfile which can be characterized as a work $I_B = (x_{B_1}, \ldots, x_{B_m})$, whose elements reside in the universe *Bit* ($\mathcal{I}_B$). Hence, a bitfile I_B with $|I_B| = m$ can also be considered as a binary sequence $I_B = \{0,1\}^m$.

Let $\mathcal{T}(\cdot)$ be a transformation, which transforms a work on a specific abstraction level into a work of another abstraction level. A transformation from the higher level Y to the lower abstraction level Z is denoted $\mathcal{T}_{Y \to Z}(\cdot)$, whereas a transformation from a lower to a higher level is denoted $\mathcal{T}_{Y \leftarrow Z}(\cdot)$.

Let $Dist_Y(\cdot, \cdot)$ be a distance function which is able to measure the differences of two works of the same abstraction level. If the distance of two IP cores I_Y and I'_Y of the same abstraction level Y is smaller than a threshold value t_I ($Dist_Y(I_Y, I'_Y) < t_I$), the two works may be considered similar.

A watermark W_Y is a vector $W_Y = (w_{Y1}, w_{Y2}, \ldots, w_{Yl})$, where each element $w_{Yi} \in \mathcal{W}_Y$. The universe $\mathcal{W}_Y$ is dependent on the universe of the work $\mathcal{I}_Y$ and the watermark generation process. A key K is a sequence of m binary bits ($K = \{0,\ 1\}^m$).

In the watermark model, there exist three algorithms: the *watermark generator* $\mathcal{G}$, the *watermark embedder* $\mathcal{E}$, and the *watermark detector* $\mathcal{D}$. In detail, a specific watermark generator $\mathcal{G}_X(\cdot)$ is able to generate a watermark W_X for the abstraction level X from a key K: $W_X = \mathcal{G}_X(K)$. The input of the watermark embedder or detector must be in the same abstraction level. For example, to watermark an IP core I_X at abstraction level X, also the watermark W_X must be generated for this abstraction level. So to obtain a watermarked work on the abstraction level X, it is necessary to also use a watermarked and an embedding algorithm suitable for the same abstraction level, i.e., $\widetilde{I}_X = \mathcal{E}_X(I_X, W_X)$. The watermark in $\widetilde{I}_X$ should obviously not be visible. Therefore, the difference between I_X and $\widetilde{I}_X$ should be small. With the distance function, this can be expressed as $Dist_X(I_X, \widetilde{I}_X) < t_I$, where t_I is a threshold value upon the difference is noticeable. Using the watermark detector $\mathcal{D}_X$, the existence of the watermark W_X in the work $\widetilde{I}_X$ can be proven, if $\mathcal{D}_X(\widetilde{I}_X, W_X) = true$ or negated if $\mathcal{D}_X(\widetilde{I}_X, W_X) = false$.

In order to achieve full transparency of the watermarking process towards design tools, it is an essential requirement that a work, marked on any abstraction level, will retain the watermark if transformed to a lower abstraction level. Hence, if $\mathcal{D}_Y(\widetilde{I}_Y, W_Y) = true$, so should also $\mathcal{D}(\widetilde{I}_Z, W_Z) = true$, if $\widetilde{I}_Z = \mathcal{T}_{Y \to Z}(\widetilde{I}_Y)$, and W_Z is a representation of W_Y on abstraction level Z.

However, considering reverse engineering, the watermark information may be removed by the reverse engineering transformation $\mathcal{T}_{Y \leftarrow Z}(\cdot)$, or the detection and removal of the watermark may be greatly simplified on the higher abstraction level. For example, consider an FPGA bitfile IP core watermarking technique for which the watermark is stored in some placement information inside the bitfile. The watermark is generated for bitfiles on the device level: $W_B = \mathcal{G}_B(K)$ and is embedded in a bitfile core I_B to create the watermarked bitfile: $\widetilde{I}_B = \mathcal{E}_B(I_B, W_B)$. If an attacker is able to reverse engineer the

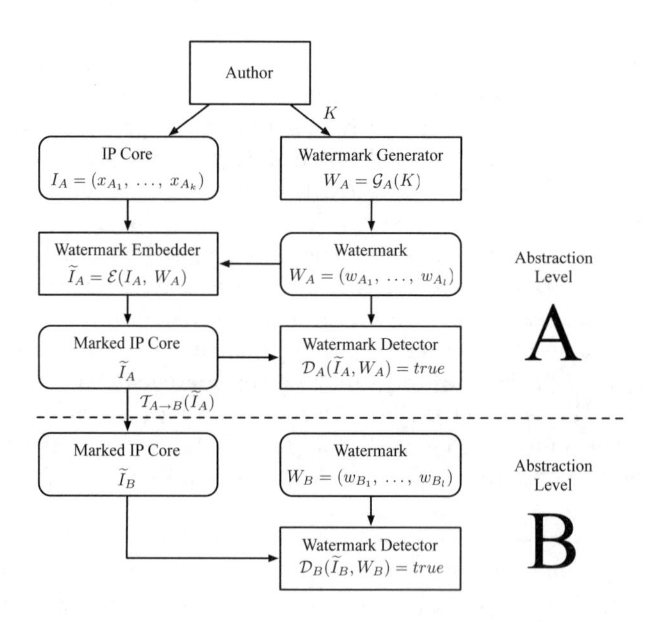

Fig. 3. An example of a watermarking procedure characterized in the IP core watermark model with different abstraction levels. At abstraction level A, the watermark is generated and embedded. A transformation to the abstraction level B retains the watermark [20].

bitfile and reconstruct a netlist on the logic level, the placement information will get lost, since there is no representation for this information on the logic level. This implies, of course, that the watermark is lost, as well: $\widetilde{I}_L = \mathcal{T}_{L \leftarrow B}(\widetilde{I}_B)$, $\mathcal{D}_L(\widetilde{I}_L, W_L) = false$. Another problem of reverse engineering may be that an embedded watermark might become obviously readable at the higher abstraction level and can be removed easily.

Figure 3 shows an example of the IP core watermark model considering different abstraction levels.

3.2 Threat Model

In the general multimedia watermarking model introduced by Li et al. [17], it should be computationally infeasible to remove the watermark without changing the properties of the work. For the introduced IP core watermarking model, this requirement does not necessarily hold. Sometimes, it might be easier for an attacker to redevelop an IP core than to remove a watermark. The question to purchase or to redevelop a core is a pure matter of cost. An uprising economical question is whether the development of an attack is an option. For many cases, the redevelopment from scratch might be cheaper than obtaining an unlicensed core and develop an attack in order to remove the watermark. On the other hand, there are designs involving such cunning cleverness and creativity that trying to redevelop a work of equivalent economic value would exceed the costs of developing an appropriate attack by several orders of magnitude.

We may consider a watermarking technique secure, if the cost for obtaining an unlicensed IP core and developing a removal attack is higher than to purchase the IP core.

Let $\mathcal{A}_Y$ be an algorithm which is able to transform a watermarked IP core $\widetilde{I}_Y$ at the abstraction level Y into an IP core with removed or disabled watermark $I'_Y = \mathcal{A}_Y(\widetilde{I}_Y)$. Let $C(\cdot)$ be a cost function. Furthermore, denote $C_D(\cdot)$ as the development cost of a specified IP core or attack and $C_P(\cdot)$ the purchase cost of an IP core. Let $C_O(\cdot)$ denote the cost to obtain an (unlicensed) IP core. Note that this cost may vary between the costs for copying the core from an arbitrary source and those for purchasing it. We define a watermarked core $\widetilde{I}_Y$ to be secure against attacks if attacks produce higher costs than the legal use of the core. Instead of requiring computational infeasibility, it is enough to fulfill:

$$C_P(\widetilde{I}_Y) < C_D(I_Y) \leq (C_O(\widetilde{I}_Y) + C_D(\mathcal{A}_Y(\widetilde{I}_Y))). \tag{1}$$

Furthermore, a reverse engineering step to a higher abstraction level and the development of an attacker algorithm on this level might be cheaper than the development of an attacker algorithm on the lower abstraction level. Therefore, we must also consider the usage of reverse engineering:

$$C_P(\widetilde{I}_Y) < C_D(I_Y) \leq (C_O(\widetilde{I}_Y) + C(\mathcal{T}_{X \leftarrow Y}(\widetilde{I}_Y)) + C_D(\mathcal{A}_X(\widetilde{I}_X)) + C(\mathcal{T}_{X \rightarrow Y}(I'_X)). \tag{2}$$

Definition 1. *An IP core watermarking scheme is called t_I-resistant to removal attacks if for any attacker $\mathcal{A}$ and any IP core $\widetilde{I}_Y$ of a given abstraction level Y and watermarked by W_Y, it is either computationally infeasible to compute $I'_Y = \mathcal{A}(\widetilde{I}_Y)$ with $Dist_Y(\widetilde{I}_Y, I'_Y) < t_I$ and $\mathcal{D}_Y(I'_Y, W_Y) = false$ or produces higher costs than its legal use.*

The term t_I*-resistant* means that the watermark scheme is resistant against removal attacks with respect to the threshold value t_I. If the distance exceeds t_I, the works cannot be counted as identical. For example, if the attacker creates a completely new work, the watermark is also removed, but the works are not the same. The phrase *computationally infeasible* follows the standard definition from cryptography. Something is computationally infeasible if the cost (e.g., memory, runtime, area) is finite but impossibly large [6]. Here, this is true if the probability $Pr[\mathcal{A}_Y(\widetilde{I}_Y) = I'_Y]$ is negligible with respect to the problem size n. A quantity X is negligible with respect to n if and only if for all sufficiently large n and any fixed polynomial $q(\cdot)$ (the attacker $\mathcal{A}_Y$ is defined as an algorithm of polynomial complexity), we have $X < 1/q(n)$ [17].

In other words, with a sufficiently large problem size of watermarked work $\widetilde{I}_Y$, resistance against removal attacks means that the attacker is unable to remove the watermark as the problem size is beyond the computational capability of the attacker, unless the resulting work is perceptually different from the original work.

For ambiguity attacks where an attacker tries to counterfeit the ownership or to achieve an ownership deadlock, the attacker searches for a fake watermark inside the IP core. This can be done by analyzing the IP core and searching for a structural or statistical feature which might be suitable to be interpreted as a fake watermark. However, the published IP core may be delivered in different target technology versions, for example, for an ASIC design flow or for different FPGA target devices. This fake watermark must of course be present in any other distributed version of the IP core in order to guarantee the attacker's authentic evidence of ownership. Furthermore, the attacker must present a fake original work and the evidence of a comprehensible watermark generation from a unique key, which clearly identifies the attacker. These are all reasons why ambiguity attacks are very difficult in the area of IP core watermarking.

Definition 2. *An IP core watermarking scheme is called resistant to ambiguity attacks if for any attacker $\mathcal{A}$ and any given IP core $\widetilde{I}_Y$ of a certain abstraction level Y and watermarked by W_Y, it is either computationally infeasible to compute a valid watermark W'_Y such that $\mathcal{D}_Y(\widetilde{I}_Y, W'_Y) = true$ or produces more costs than its legal use.*

In case of key copy attacks, the key of a credible author is used to watermark a work with lower quality. In general, it should be impossible for an attacker to create a work I'_Y which is distinguishable from any work of another author where the key or watermark of a credible author can be found.

Definition 3. *A watermarking scheme is t_I-resistant to key copy attacks if for any attacker $\mathcal{A}_Y$ and any work $\widetilde{I}_Y = \mathcal{E}_Y(I_Y, W_Y)$ for some original I_Y and the* *watermark W_Y, it is computationally infeasible for $\mathcal{A}_Y$ to compute a work I'_Y such that $Dist(I_Y, I'_Y) > t_I$, yet $\mathcal{D}_Y(I'_Y, W_Y) = true$ [17].*

To prevent key copy attacks, a private/public key algorithm, like RSA [18] can be used. RSA is an asymmetrical cryptography method that is based on factorization of a number into prime numbers. The author encrypts a message which clearly identifies the author and the work with his private key. The work can be identified by a hash value over the original work. This encrypted message is now used for generating the watermark and embedded inside the work. Stealing this watermark is useless, because everyone can decrypt the message with the public key, whereas no one can alter this message.

4 Watermark Verification Strategies for Embedded FPGAs

The problem of applying watermarking techniques to FPGA designs is not the coding and insertion of a watermark, rather it is the verification with an FPGA embedded in a system that poses the real challenge. Hence, our methods concentrate in particular on the verification of watermarks. When considering finished products, there are five potential sources of information that can be used for extracting a watermark: The configuration bitfile, the ports, the power consumption, electromagnetic (EM) radiation, and the temperature.

If the developer of an FPGA design has disabled the possibility to simply read back the bitfile from the chip, it can be extracted by wire tapping the communication between the PROM and the FPGA. Some FPGA manufactures provide an option to encrypt the bitstream which will be decrypted only during configuration inside the FPGA. Monitoring the communication between PROM and FPGA in this case is useless, because only the encrypted file will be transmitted. Configuration bitfiles mostly use a proprietary format which is not documented by the FPGA manufacturers. However, it seems to be possible to read out some parts of the bitfile, such as information stored in RAMs or lookup tables. In Section 5, we introduce a procedure in which the watermarks are inserted into an IP core specified on the logic level in form of a netlist and can then be extracted from the configuration bitstream.

Another popular approach for retrieving a signature from an FPGA is to employ unused ports. Although this method is applicable to top-level designs, it is impractical for IP cores, since these are mostly used as components that will be combined with other resources and embedded into a design so that the ports will not be directly accessible any more. Due to these restrictions, we do not discuss the extraction of watermarks over output ports.

Furthermore, it is possible to force patterns on the power consumption of an FPGA, which can be used as a covert channel to transmit data to the outside of the FPGA. We have shown in [27] and [24] that the clock frequency and toggling logic can be used to control such a power spectrum covert channel. The basic idea to use these techniques for watermarking is to force a signature dependent toggle pattern and extract the resulting change in power consumption as a signature from the FPGA's power spectrum. We refer to this method as "Power Watermarking" in Section 6.

With almost the same strategy it is also possible to extract signatures from the electro magnetic (EM) radiation of an FPGA. A further advantage of this technique is that a raster scan of an FPGA surface with an EM sensor can also use the location information to extract and verify the watermark. Unfortunately, more and more FPGAs are delivered in a metal chip package which absorbs the EM radiation. Nevertheless, this is an interesting alternative technique for extracting watermarks and invites for future research.

Finally, a watermark might be read out by monitoring the temperature radiation. The concept is similar to the power and EM-field watermarking approaches, however, the transmission speed is drastically reduced. Interestingly, this is the only watermarking approach which is commercially available [11]. Here, reading the watermark from an FPGA may take up to 10 minutes.

5 Watermark Verification Using the FPGA Bitfile

In this section we present a method where an embedded signature is extracted from an FPGA bitfile. We start out by discussing how the contents of the lookup tables may be extracted from the FPGA bitfile. Following, a watermarking method for netlist cores is proposed (see also [20]).

5.1 Lookup Table Content Extraction

In order to harden the watermark against removal it is very important to integrate the watermark into the functional parts of the IP core, so that simply removing the mark carrying components would damage the core. For FPGA designs, the functional lookup tables are an ideally suited component for carrying watermarks. From a finished product, it is possible to obtain the configuration bitstream of the FPGA. The extraction of the lookup table contents from the configuration bitfile depends on the FPGA device and the FPGA vendor. To read out the LUT content directly from the bitfile, it must be known at which position in the bitfile the lookup table content is stored and how these values must be interpreted. In [22], for example, a standard black-box reverse engineering procedure is applied to interpret Xilinx Virtex-II and Virtex-II Pro bitfiles. To generalize this approach, we define a lookup table extractor function $\mathcal{L}_X(\cdot)$ for the abstraction level X. The extractor function is able to extract the lookup table content of a work I_X as follows: $\mathcal{L}_X(I_X) = \{x_{X_1}, x_{X_2}, \ldots, x_{X_m}\}$, whereas x_{X_i} is a lookup table content element of the abstraction level X, and m is the number of used lookup tables. Here it is essential to include the abstraction level, because LUT content might be encoded differently on different technology representations. The extraction function can be applied to extract the lookup table contents of a design I_B of the bitfile on abstraction level

B: $\mathcal{L}_B(I_B) = \{x_{B_1}, x_{B_2}, \ldots, x_{B_q}\}$. Each element x_{B_i} consists of the lookup table content as well as the slice coordinates of the corresponding lookup table. Once again, we have to mention that these methods will not work with bitfile encryption.

5.2 Watermarks in Functional LUTs for Netlist Cores

Since we want to keep the IP core as versatile as possible, we watermark the design in the form of a netlist representation, which, although technology dependent to a certain degree, can still be used for a large number of different devices. Netlist designs will almost certainly undergo the typical design flow for silicon implementations. This also includes very sophisticated optimization algorithms, which will eliminate any redundancy that can be found in the design in order to make improvements. As a consequence it is necessary to embed the watermarks in the netlist in such a way, that the optimization tools will not remove the watermarks from the design. In Xilinx FPGAs, for example, lookup tables are essentially RAM cells, with the inputs specifying which of the stored bits to deliver to the output of the RAM. Naturally, these cells can therefore also be used as storage, but also as shift-register cells (see Figure 4). Interesting, however, is the fact that if the cell is configured as a lookup table, Xilinx optimization tools will try to optimize the contained logic function. If the cell is in contrast configured as a shift-register or distributed RAM, the optimization tools will leave the contents alone, but the logic function is still carried out. This means, that if we want to add redundancy to a netlist, that is not removed by automized tools, all we have to do is to take the corresponding cells out of the scope of the tools. FPGAs usually consist of the same type of lookup tables with respect to the number of inputs. For example, the Xilinx Virtex-II uses lookup tables with four inputs whereas the Virtex-5 has lookup tables with six inputs. However, in common netlist cores many logical lookup tables exist, which have less inputs than the type used on the FPGA. These lookup tables are mapped to the physical lookup tables of the FPGA during synthesis. If the logical lookup table of the netlist core has fewer inputs than the physical representation, the memory space which was not present in the logical representation remains unused. Using the unused memory space of functional lookup tables for watermarking without converting the lookup table either to a shift register or distributed memory turns out to be not applicable, because design flow tools identify the watermark as redundant and remove the content due to optimization. Converting the watermarked functional lookup table into a shift register or a memory cell prevents the watermark from deletion due to optimization.

Embedding the Watermark. The first step of embedding a watermark is to extract all lookup tables from a given netlist core I_L: $\mathcal{L}_L(I_L) = \{lut_{L_1}, lut_{L_2}, \ldots, lut_{L_r}\}$, where L denotes the logic abstraction level used for netlist cores (see Figure 5) and subscript r refers to the amount of extracted LUTs. Each element lut_{L_i} denotes a lookup table primitive cell in the netlist (e.g. for Virtex-II devices, LUT1, LUT2, LUT3, or LUT4). A watermark generator $\mathcal{G}_L(\cdot,\cdot)$ must know the

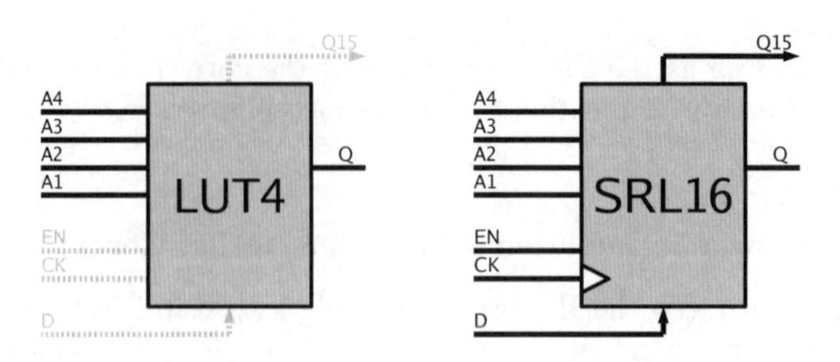

Fig. 4. In the Xilinx Virtex architecture, the same standard cell is used as a lookup table (LUT4) and also as a 16-bit shift-register lookup table(SRL16)

different lookup table cells with the functional content as well as the unique key K to generate the watermarks: $\mathcal{G}_L(K, \mathcal{L}_L(I_L)) = W_L$.

From the unique key K a secure pseudo random sequence is generated. Some or all of the extracted lookup table primitive cells are chosen to carry a watermark. Note that only lookup tables from the netlist core can be chosen which use less inputs than the physical lookup tables on the FPGA. Usually a core which is worth to be watermarked consists of many markable lookup tables. Now, the lookup tables are transformed to shift registers, ordered, and the first 4 bits of the free space are used for a counter value. The other bits are initialized according to the position with values from the pseudo random stream, generated from the key K. Note that the number of bits which can be used for the random stream depends on the original functional lookup table type.

The generated watermark W_L consists of the transformed shift registers: $W_L = \{srl_{L_1}, srl_{L_2}, \ldots, srl_{L_k}\}$ with $k \leq r$. The watermark embedder $\mathcal{E}_L$ inserts the watermarks into the netlist core I_L by replacing the corresponding original functional lookup tables with the shift registers: $\mathcal{E}_L(I_L, W_L) = \widetilde{I}_L$. The watermarked work $\widetilde{I}_L$ can now be published and sold.

Extraction of the Watermark. The purchased core $\widetilde{I}_L$ can be combined by a product developer with other purchased or self developed cores and implemented into an FPGA bitfile: $\widehat{I}_B = \mathcal{T}_{L\rightarrow B}(\widetilde{I}_L \circ I'_{L_1} \circ I'_{L_2} \circ \ldots)$ (see Figure 5). An FPGA which is programmed with this bitfile $\widehat{I}_B$ may be part of a product. If the product developer is accused of using an unlicensed core, the product can be purchased and the bitfile can be read out, e.g., by wire tapping. The lookup table content and the content of the shift registers can be extracted from the bitfile: $\mathcal{L}_B(\widehat{I}_B) = \{\widehat{x}_{B_1}, \widehat{x}_{B_2}, \ldots, \widehat{x}_{B_q}\}$.

The lookup table or shift register elements x_{B_i} belong to the device abstraction level B. The representation can differ from the representation of the same content in the logic abstraction level L. For example, in Xilinx Virtex-II FPGAs the encoding of the shift register differs from the encoding of lookup tables. For shift registers the bit order is reversed compared to the lookup table encodings. Therefore, the bitfile elements must be transferred to the logic level

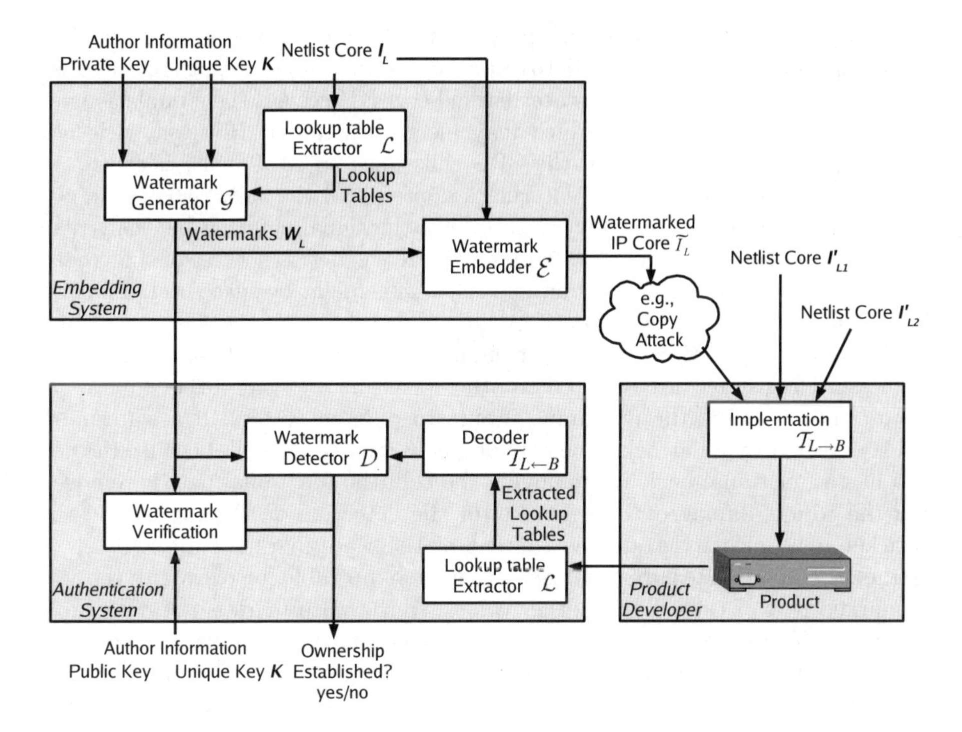

Fig. 5. The netlist core watermarking system. The embedding system is responsible for extracting the lookup tables from the netlist core, selecting suitable locations and embedding the watermarks in those LUTs that were converted into shift-registers. A product developer may obtain such a watermarked netlist core and combine it with other cores into a product. The lookup tables from the product can be extracted and transformed so that the detector can decide if the watermark is present or not.

by the corresponding decoding. This can be done by the *reverse engineering operator*: $\mathcal{T}_{L\leftarrow B}(\mathcal{L}_B(\widehat{I}_B)) = \{\widehat{x}_{L_1}, \widehat{x}_{L_2}, \ldots, \widehat{x}_{L_q}\}$. Reverse engineering lookup table or shift register content is however very simple compared to reverse engineering the whole bitfile. Now, the lookup table or shift register content can be used for the watermark detector $\mathcal{D}_L$ which can decide if the watermark W_L is embedded in the work or not: $\mathcal{D}_L(W_L, \{\widehat{x}_{L_1}, \widehat{x}_{L_2}, \ldots, \widehat{x}_{L_q}\}) = true/false$.

Robustness Analysis. To recall Section 3, the most important attacks are *removal, ambiguity, key copy,* and *copy attacks.* As stated before, a possible protection against copy attacks does not exist and key copy attacks can be prevented by using an asymmetric cryptographic method, like RSA.

Removal attacks most likely occur on the logic level, after obtaining the unlicensed core and before the integration with other cores. The first step of a removal attack is the detection of the watermarks. The appearance of the shift register primitive cells (here SRL16) in a netlist core is not suspicious because shift registers appear also in unwatermarked cores. However, the cumulative appearance may be suspicious, which may alert an attacker. However, it is not sufficient to simply convert all the SRL cells back to LUT cells, because, although this might remove the watermark content, all the shift registers in the design will be disabled. In contrast to bitfiles, the signal nets can be easily read out from a netlist core. An attacker may analyze the net structures of shift registers in order to detect the watermarked cells. This might be successful, however, we can better hide the watermark if we alter the encoding of the watermark and, therefore, the connections to the watermark cell. The reachable functional part of the shift register can be shifted to other positions by using other functional inputs and clamping the remaining inputs to different values. If a watermark cell is detected by an attacker, he cannot easily remove the cell, because the cell also has a functional part. By removing the cell, the functional part is removed and the core is damaged. Therefore, after the detection of the watermark, the attacker must either decode the content of the watermarked shift register to extract the functional part and insert a new lookup table, or overwrite the watermarked part of the cell with other values, so the watermark is not detectable any more. The different encodings of the functional part of the shift register content complicates the analysis and the extraction of it. Furthermore, even if some watermarks are removed, the establishment of the right ownership of the core is still possible, because we need not all watermarked cells for a successful detection of the signature.

In case of *ambiguity attacks*, an attacker analyzes the bitfile or the netlist to find shift register or lookup table contents which may be suitable to build a fake watermark. However, the attacker must also present the insertion procedure to achieve a meaningful result. Due to the usage of secure one way cryptographic functions for generating the watermark, the probability of a success is very low. Furthermore, the attacker can use a self-written netlist core which he watermarked with his signatures and combine it with the obtained unlicensed core. The result is, that the watermarks of the authors of both cores are found in the bitfile, which are both trustful. Inside the unique key K, not only the author information should be included but also information of the core, e.g., a hash value over the netlist core file without the watermark. Of course, the attacker can use the identification of the obtained unlicensed core for watermarking his core. However, to generate a hash value of the obtained core without watermarks, he must first remove the marks. In general, attacks against this approach are possible, but they need a high amount of effort. To increase the security against ambiguity attacks, the core may be registered at a trusted third party.

6 Power Watermarking

This section describes watermarking techniques introduced in [27] and [24], where a signature is verified over the *power consumption pattern* of an FPGA.

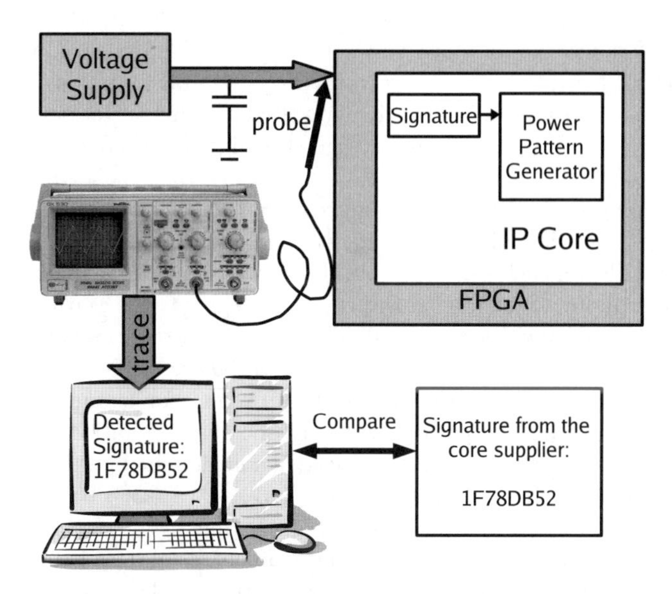

Fig. 6. Watermark verification using power signature analysis: From a signature (watermark), a power pattern inside the core will be generated that can be probed at the voltage supply pins of the FPGA. From the trace, a detection algorithm verifies the existence of the watermark.

The presented idea is new and differs from [14] and [3] where the goal of using power analysis techniques is the detection of cryptographic keys and other security issues. For power watermarking methods, the term *signature* refers to the part of the watermark which can be extracted and is needed for the detection and verification of the watermark. The signature is usually a bit sequence which is derived from the unique key for author and core identification.

There is no way to measure the relative power consumption of an FPGA directly. Only by measuring the relative supply voltage or current the actual power consumtion can be inferred. We have decided to measure the voltage of the core as close as possible to the voltage supply pins such that the smoothing from the plane and block capacities are minimal and no shunt is required. Most FPGAs have *ball grid array* (BGA) packages and the majority of them have vias to the back of the PCB for the supply voltage pins. So, the voltage can be measured on the rear side of the PCB using an oscilloscope. The voltage can be sampled using a standard oscilloscope, and analyzed and decoded using a program developed to run on a PC. The decoded signature can be compared with the original signature and thus, the watermark can be verified. This method has the advantage of being non-destructive and requires no further information or aids than the given product (see Figure 6).

In the power watermarking approach described in [26] and [27], the amplitude of the interferences in the core voltage is altered. The basic idea is to add a power pattern generator (e.g., a set of shift registers) and clock it either with the operational clock or an integer division thereof. This power pattern generator is controlled according to the encoding of the signature sequence which should be sent.

The mapping of a signature sequence $s = \{0,1\}^n$ onto a sequence of symbols $\{\sigma_0, \sigma_1\}^n$ [24] is called encoding: $\{0,1\}^n \rightarrow \mathcal{Z}^n, n \geq 0$ with the alphabet $\mathcal{Z} = \{\sigma_0, \sigma_1\}$. Here, each signature bit $\{0,1\}$ is assigned to a symbol. Each symbol σ_i is a triple $(e_i, \delta_i, \omega_i)$, with the *event* $e_i \in \{\gamma, \bar{\gamma}\}$, the *period length* $\delta_i > 0$, and the *number of repetitions* $\omega_i > 0$. The event γ is *power consumption through a shift operation* and the inverse event $\bar{\gamma}$ is *no power consumption*. The period length is given in terms of number of clock cycles. For example, the encoding through 32 shifts with the period length 1 (one shift operation per cycle) if the data bit '1' should be sent, and 32 cycles without a shift operation for the data bit '0' is defined by the alphabet $\mathcal{Z} = \{(\gamma, 1, 32), (\bar{\gamma}, 1, 32)\}$.

Different power watermarking encoding schemes were introduced and analyzed. The basic method with encoding scheme: $\mathcal{Z} = \{(\gamma, 1, 1), (\bar{\gamma}, 1, 1)\}$, the enhanced robustness encoding: $\mathcal{Z} = \{(\gamma, 1, 32), (\bar{\gamma}, 1, 32)\}$, and the BPSK approach: $\mathcal{Z} = \{(\gamma, 1, \omega), (\bar{\gamma}, 1, \omega)\}$ are explained in detail in [27]. The correlation method with encoding $\mathcal{Z} = \{(\gamma, 25, 1), (\bar{\gamma}, 25, 1)\}$ can be reviewed in [24]. To avoid interference from the operational logic in the measured voltage, the signature is only generated during the reset phase of the core.

The power pattern generator consists of several shift registers, causing a recognizable signature- and encoding-dependent power consumption pattern. As mentioned before in Section 5.2, a shift register can also be used as a lookup table and vice versa in many FPGA architectures (see Figure 4 in Section 5.2). A conversion of functional lookup tables into shift registers does not affect the functionality if the new inputs are set correctly. This allows us to use functional logic for implementing the power pattern generator. The core operates in two modes, the *functional mode* and the *reset mode*. In the functional mode, the shift is disabled and the shift register operates as a normal lookup table. In the reset mode, the content is shifted according to the signature bits and consumes power which can be measured outside of the FPGA. To prevent the loss of the content of the lookup table, the output of the shift register is fed back to the input, such that the content is shifted circularly. When the core changes to the functional mode, the content have to be shifted to the proper position to get a functional lookup table for the core.

To increase the robustness against removal and ambiguity attacks, the content of the power consumption shift register which is also part of the functional logic can be initialized shifted. Only during the reset state, when the signature is transmitted, the content of the functional lookup table can be positioned correctly. So, normal core operation cannot start before the signature was transmitted completely. The advantage is that the core is only able to work after sending the signature. Furthermore, to avoid a too short reset time in which the watermark cannot be detected exactly, the right functionality will only be established if the reset state is longer than a predefined time. This prevents the user from leaving out or shorten the reset state with the result that the signature cannot be detected properly.

The signature itself can be implemented as a part of the functional logic in the same way. Some lookup tables are connected together and the content, the function of the LUTs, represents the signature. Furthermore, techniques described in Section 5.2 can be used to combine an additional watermark and the functional part in a single lookup table if not all lookup table inputs are used for the function. For example, LUT2 primitives in Xilinx Virtex-II devices can be used to carry an additional 12-bit watermark by restricting the reachability of the functional lookup table through clamping certain signals to constant values. Therefore, the final sending sequence consists of the functional part and the additional watermark. This principle makes it almost impossible for an attacker to change the content of the signature shift register. Altering the signature would also affect the functional core and thus result in a corrupt core.

The advantages of using the functional logic of the core as a shift register are the reduced resource overhead for watermarking and the robustness of this method. It is hard, if not impossible, to remove shift registers without destroying the functional core, because they are embedded in the functional design. The watermark embedder $\mathcal{E}_L(I_L, W_L) = \widetilde{I}_L$ consists of two steps. First, the core I_L must be embedded in a wrapper which contains the control logic for emitting the signature. This step is done at the register-transfer level before synthesis. The second step is at the logic level after the synthesis. A program converts suitable lookup tables (for example LUT4 for Virtex-II FPGAs) into shift registers for the generation of the power pattern and attaches the corresponding control signal from the control logic in the wrapper. The wrapper contains the control logic for emitting the watermark and a register that contains the signature. The ports of the wrapper are identical to the core, so we can easily integrate this wrapper into the hierarchy. The control logic enables the signature register while the core is in reset state. Also, the power pattern shift registers are shifted in correspondence to the current signature bit. If the reset input of the wrapper is deasserted, the core function cannot start immediately, but only as soon as the content in the shift registers has been shifted back to the correct position. Then the control logic deasserts the internal reset signal to enter normal function mode. The translation of four input lookup tables (LUT4) of the functional logic into 16 Bit shift registers (SRL16) is done at the netlist level. The watermarked core $\widetilde{I}_L$ is now ready for purchase or publication. A company may obtain an unlicensed version of the core $\widehat{I}_L$ and embeds this core in a product: $\widehat{I}_P = \mathcal{T}_{L \to B}(\widehat{I}_L \circ I'_{L_1} \circ I'_{L_2} \circ \ldots)$. If the core developer has a suspicious fact, he can buy the product and verify that his signature is inside the core using a detection function $\mathcal{D}_P(\widehat{I}_P, W_L) = true/false$. The detecting function depends on the encoding scheme. In [27] and [24], the detecting functions of all introduced encoding schemes are described in detail.

The advantage of power watermarking is that the signature can easily be read out from a given device. Only the core voltage of the FPGA must be measured and recorded. No bitfile is required which needs to be reverse-engineered. Also, these methods work for encrypted bitfiles where methods extracting the signature from the bitfile fail. Moreover, we are able to sign netlist cores, because our watermarking algorithm does not need any placement information. However, many watermarked netlist cores can be integrated into one design. The results are superpositions and interferences which complicate or even prohibit the correct decoding of the signatures. To achieve the correct decoding of all signatures, we proposed *multiplexing* methods in [23].

Robustness Analysis. The most common attacks against watermarking mentioned in Section 3 are *removal, ambiguity, key copy*, and *copy attacks*. Once again, key copy attacks can be prevented by asymmetric cryptographic methods, and there is no protection against copy attacks.

Removal attacks most likely take place on the logic level instead of the device level where it is really hard to alter the design. The signature and power shift registers as well as the watermark sending control logic in the wrapper are mixed with functional elements in the netlist. Therefore, they are not easy to detect. Even if an attacker is able to identify the sending logic, a deactivation is useless if the content of the power shift register is only shifted into correct positions after sending the signature. By preventing the sending of the watermark, the core is unable to start. Another possibility is to alter the signature inside the shift register. The attacker may analyze the netlist to find the place were the signature is stored. This attack is only successful if there is no functional logic part mixed with the signature. By mixing the random bits with functional bits, it is hard to alter the signature without destroying the correct functionality of the core. Therefore, this watermark technique can be considered as resistant against removal attacks.

In case of *ambiguity attacks*, an attacker analyses the power consumption of the FPGA in order to find a fake watermark, or to implement a core whose power pattern disturbs the detection of the watermark. In order to trustfully fake watermarks inside the power consumption signal, the attacker must present the insertion and sending procedure which should be impossible without using an additional core. Another possibility for the attacker is to implement a *disturbance core* which needs a lot of power and makes the detection of the watermark impossible. In [27] and [24], *enhanced robustness encoding methods* are presented which increase the possibility to decode the signature, even if other cores are operating during the sending of the signature. Although a disturbance core might be successful, this core needs area and most notably power which increases the costs for the product. The presence of a disturbance core in a product is also suspicious and might lead to further investigation if a copyright infringement has occurred. Finally, the attacker may watermark another core with his watermark and claim that all cores belong to him. This can be prevented by adding a hash value of the original core without the watermark to the signature like in the bitfile watermarking method for netlist cores. The sending of watermarks of multiple cores at the same time is addressed in [23].

7 Conclusions

In this paper, we have presented exemplary two different approaches for watermarking of IP cores. Our methods follow the strategy of an easy verification of the watermark or the identification of the core in a bought product from an accused company without any further information. Netlist cores, which have a high trade potential for embedded systems developers, are in the focus of our analysis. To establish the authorship in a bought product by watermarking or core identification, we have discovered different new techniques, how information can be transmitted from the embedded core to the outer world. In this paper, we concentrated on methods using the *FPGA bitfile* which can be extracted from the product and on methods where the signature is transmitted over the *power pins* of the FPGA. In Section 3, we adapt the *theoretical general watermark approach* from Li et al. [17] for IP core watermarking and show possible threats and attacks. Section 5 deals with IP core watermarking methods where the authorship is established by analysis of the extracted bitfile. In Section 6, we have described watermark techniques for IP cores where the signature can be extracted easily over the power pins of the chip. The main idea is that during a reset phase of a chip, a watermark circuit is responsible to emit a characteristic power pattern sequence that may be measured by voltage fluctuations on power pins. With these techniques, it is possible to decide with high confidence, whether an IP core of a certain vendor is present on the FPGA or not. For all methods, we analyzed the strengths and weaknesses in case of removal of ambiguity attacks.

References

1. Abdel-Hamid, A.T., Tahar, S., Aboulhamid, E.M.: A Survey on IP Watermarking Techniques. Design Automation for Embedded Systems 9(3), 211–227 (2004)
2. Adelsbach, A., Rohe, M., Sadeghi, A.-R.: Overcoming the Obstacles of Zero-knowledge Watermark Detection. In: Proceedings of the 2004 Workshop on Multimedia and Security MM&Sec 2004, pp. 46–55. ACM, New York (2004)
3. Agrawal, D., Archambeault, B., Rao, J.R., Rohatgi, P.: The EM Side-Channel(s). In: Kaliski Jr., B.S., Koç, Ç.K., Paar, C. (eds.) CHES 2002. LNCS, vol. 2523, pp. 29–45. Springer, Heidelberg (2003)
4. Bai, F., Gao, Z., Xu, Y., Cai, X.: A Watermarking Technique for Hard IP Protection in Full-custom IC Design. In: International Conference on Communications, Circuits and Systems (ICCCAS 2007), pp. 1177–1180 (2007)
5. Boney, L., Tewfik, A.H., Hamdy, K.N.: Digital Watermarks for Audio Signals. In: International Conference on Multimedia Computing and Systems, pp. 473–480 (1996)

6. Diffie, W., Hellman, M.E.: New Directions in Cryptography. IEEE Transactions on Information Theory 22(6), 644–654 (1976)
7. Drimer, S.: Security for Volatile FPGAs (November 2009)
8. Fan, Y.C., Tsao, H.W.: Watermarking for Intellectual Property Protection. Electronics Letters 39(18), 1316–1318 (2003)
9. Kahng, A.B., Lach, J., Mangione-Smith, W.H., Mantik, S., Markov, I.L., Potkonjak, M.M., Tucker, P.A., Wang, H., Wolfe, G.: Constraint-Based Watermarking Techniques for Design IP Protection. IEEE Transactions on Computer-Aided Design of Integrated Circuits and Systems 20(10), 1236–1252 (2001)
10. Kahng, A.B., Mantik, S., Markov, I.L., Potkonjak, M.M., Tucker, P.A., Wang, H., Wolfe, G.: Robust IP Watermarking Methodologies for Physical Design. In: Proceedings of the 35th Annual Design Automation Conference DAC 1998, pp. 782–787. ACM, New York (1998)
11. Kean, T., McLaren, D., Marsh, C.: Verifying the Authenticity of Chip Designs with the DesignTag System. In: Proceedings of the 2008 IEEE International Workshop on Hardware-Oriented Security and Trust HOST 2008, Washington, DC, USA, pp. 59–64. IEEE Computer Society, Los Alamitos (2008)
12. Kirovski, D., Hwang, Y.-Y., Potkonjak, M., Cong, J.: Intellectual Property Protection by Watermarking Combinational Logic Synthesis Solutions. In: Proceedings of the 1998 IEEE/ACM International Conference on Computer-Aided Design ICCAD 1998, pp. 194–198. ACM, New York (1998)
13. Kirovski, D., Potkonjak, M.: Intellectual Property Protection Using Watermarking Partial Scan Chains For Sequential Logic Test Generation. In: Proceedings of the 1998 IEEE/ACM International Conference on Computer-Aided Design ICCAD 1998 (1998)
14. Kocher, P.C., Jaffe, J., Jun, B.: Differential Power Analysis. In: Wiener, M. (ed.) CRYPTO 1999. LNCS, vol. 1666, pp. 388–397. Springer, Heidelberg (1999)
15. Lach, J., Mangione-Smith, W.H., Potkonjak, M.: Signature Hiding Techniques for FPGA Intellectual Property Protection. In: Proceedings of the 1998 IEEE/ACM International Conference on Computer-Aided Design ICCAD 1998, pp. 186–189. ACM, New York (1998)
16. Li, Q., Chang, E.-C.: Zero-knowledge Watermark Detection Resistant to Ambiguity Attacks. In: Proceedings of the 8th Workshop on Multimedia and Security MMSec 2006, pp. 158–163. ACM, New York (2006)
17. Li, Q., Memon, N., Sencar, H.T.: Security Issues in Watermarking Applications – A Deeper Look. In: Proceedings of the 4th ACM International Workshop on Contents Protection and Security MCPS 2006, pp. 23–28. ACM, New York (2006)
18. Rivest, R.L., Shamir, A., Adleman, L.M.: A Method for Obtaining Digital Signatures and Public-key Cryptosystems. Communications of the ACM 21(2), 120–126 (1978)
19. Saha, D., Sur-Kolay, S.: Fast Robust Intellectual Property Protection for VLSI Physical Design. In: Proceedings of the 10th International Conference on Information Technology ICIT 2007, Washington, DC, USA, pp. 1–6. IEEE Computer Society, Los Alamitos (2007)
20. Schmid, M., Ziener, D., Teich, J.: Netlist-Level IP Protection by Watermarking for LUT-Based FPGAs. In: Proceedings of IEEE International Conference on Field-Programmable Technology (FPT 2008), Taipei, Taiwan, pp. 209–216 (December 2008)

21. Xilinx Inc. JBits 3.0 SDK for Virtex-II, http://www.xilinx.com/labs/projects/jbits/
22. Ziener, D., Aßmus, S., Teich, J.: Identifying FPGA IP-Cores based on Lookup Table Content Analysis. In: Proceedings of 16th International Conference on Field Programmable Logic and Applications (FPL 2006), Madrid, Spain, pp. 481–486 (August 2006)
23. Ziener, D., Baueregger, F., Teich, J.: Multiplexing Methods for Power Watermarking. In: Proceedings of the IEEE Int. Symposium on Hardware-Oriented Security and Trust (HOST 2010), Anaheim, USA (June 2010)
24. Ziener, D., Baueregger, F., Teich, J.: Using the Power Side Channel of FPGAs for Communication. In: Proceedings of the 18th Annual International IEEE Symposium on Field-Programmable Custom Computing Machines (FCCM 2010), pp. 237–244 (May 2010)
25. Ziener, D., Teich, J.: Evaluation of Watermarking Methods for FPGA-Based IP-cores. Technical Report 01-2005, University of Erlangen-Nuremberg, Department of CS 12, Hardware-Software-Co-Design, Am Weichselgarten 3, D-91058 Erlangen, Germany (March 2005)
26. Ziener, D., Teich, J.: FPGA Core Watermarking Based on Power Signature Analysis. In: Proceedings of IEEE International Conference on Field-Programmable Technology (FPT 2006), Bangkok, Thailand, pp. 205–212 (December 2006)
27. Ziener, D., Teich, J.: Power Signature Watermarking of IP Cores for FPGAs. Journal of Signal Processing Systems 51(1), 123–136 (2008)

Efficient and Flexible Co-processor for Server-Based Public Key Cryptography Applications

Ralf Laue

KOBIL Systems GmbH
67547 Worms, Germany
ralf.laue@kobil.com

Abstract. This work presents a SoC-based co-processor for public key cryptography and server application. Because of the focus on the server side, high throughput was chosen as metric for efficiency instead of low latency as usually done in literature. This becomes important in light of the second goal, which is flexibility regarding the supported cryptographic schemes. Starting with an unified view on the abstraction levels of different public key cryptographic schemes and an overview on their parallelization possibilities, parallelization is applied in a more refined way than usually done in literature: It is examined on each abstraction level which degree of parallelization still aids throughput without sacrificing flexibility.

Keywords: Public Key Cryptography, Parallelization, System-on-a-Chip.

1 Introduction

Client-server networks become increasingly heterogeneous concerning the involved platforms and protocols. As secure channels are becoming more important, cryptography is included into many protocols. For the server, which has to execute cryptographic operation for each client, this introduces a considerable computational overhead. This overhead is especially significant for *Public Key Cryptography*, which is needed to solve the problem of securely distributing keys.

To help servers shoulder this additional computational load, this work proposes a cryptographic co-processor, which encapsulates all the necessary operations as a System-on-a-Chip (SoC). Additionally, a server using different protocols for communication, may need to support different cryptographic algorithms. Therefore, such a co-processor should be designed with two main goals:

Efficiency. To enable the server to communicate with as many clients as possible, it should exhibit a high throughput, i.e., number of cryptographic operations per second. Low latency is not as important, as long as the cryptographic operation does not become the bottleneck in the overall communication process.

A. Biedermann and H. Gregor Molter (Eds.): Secure Embedded Systems, LNEE 78, pp. 129–149.
springerlink.com

Flexibility. The co-processor should be flexible concerning the supported cryptographic algorithms. This also means that it should be relatively easy to add support of new cryptographic protocols.

Implementing the cryptographic operations on a SoC co-processor also aids reusability and helps security: The reusability is increased, as the co-processor incorporates all required functions, which simplifies the control tasks for the host server. The security is helped, because the secret information does not need to leave the physical chip, which considerably decreases the eavesdropping risk.

Today, the most common public key scheme is RSA [29]. Although it features a high trust in its security, because of the amount of time it is known, there are many alternative proposals. The most important is *Elliptic Curve Cryptography* (ECC) [9]. Although it is more complicated, its computational complexity increases slower for higher security levels than that of RSA. Additional, this work considers *Pairing Based Cryptography* (PBC) [8], which hasn't fully outgrown its research phase yet.

To gain a design exhibiting the desired properties efficiency and flexibility, this work presents an abstraction level framework highlighting similarities and differences of the different algorithms of RSA, ECC, and PBC. Based on this, an architecture is derived, which uses as much parallelization on each abstraction level as it still provides throughput gains for all supported algorithms.

To evaluate the validity of the chosen approach an SoC-based prototype implementation of this architecture was realized on an FPGA. In short, FPGAs are reprogrammable hardware units. In terms of speed and development effort they lie between pure hardware and software implementations. Because of their reprogrammability, they are well-suited for prototype implementations. For a more detailed overview the reader is referred to [36].

The remainder of this work is structured as follows: The next section contains an overview on literature with similarities to this work. Section 3 gives more details on the considered cryptographic algorithms and shows the derived abstraction level framework. It also provides an overview on parallelization possibilities for public key cryptography and explains, which options were chosen on the different levels. The derived co-processor architecture is presented in Section 4 also stating the design decisions leading to it. Section 5 shows a prototype implementation of this architecture. The results gained from this implementation are presented in Section 6 together with a comparison with results from literature. Section 7 concludes this work by summarizing the important points.

2 Related Work

Related work may be divided into the aspects *RSA/ECC combinations*, *cryptographic HW/SW combinations*, and *single-chip solutions*. Note that hardware designs for PBC are not considered here, because such implementations from literature concentrate on $\mathbb{GF}(2^m)$.

Although many designs from literature notice the similarities between RSA and ECC and consider them both, most must be instantiated for a certain bit-width and do not aim for equal support of both, see [24,26,23,2]. [2] somewhat addresses this problem by extending an existing RSA implementation into an ECC implementation by reusing the modular multiplier and adding parts like other modular operations and a module executing the EC operations. However, this approach does not lead to a continuous exploitation of all resources, as the larger part of the modular multiplier will stay unused for ECC. A design combining support for RSA and ECC while trying to minimize the amount of unused resources is presented in [7]. It proposes the use of several smaller modular multipliers optimized for the bit-widths of ECC, which may either be used independently in parallel or combined together into a single larger multiplier to compute numbers of larger bit-widths for RSA. This still leads to unused resources for bit-widths, which are not a multiple of the bit-width of the smaller cores and requires the possibility to schedule ECC and RSA operations.

HW/SW co-design allows implementations, which execute the parts of an algorithm either in hardware or software depending on where it is most advantageous. Such designs for cryptography may be found in [22,32], both realizing the modular multiplication in hardware and the exponentiation needed for RSA in software. An HW/SW-realization concentrating on ECC may be found in [31].

[37,13,25] present cryptographic single-chip solutions consisting of a single general purpose processor core and dedicated cores for modular arithmetic and auxiliary functions. The design from [13] features a general purpose processor and cores for RSA, ECC, AES, and hash function. However, the possible synergies between RSA and ECC are not exploited and the cores are optimized for a single bit-width for both RSA and ECC, making it less flexible and upgradeable than this work. [25] has a general purpose processor and cores for symmetric encryption, RSA/ECC, random number generation, and MPEG en-/decoding. The authors' aim was the presentation of a new design methodology using cryptographic cores to examine their technique. Therefore, [25] does not focus on achieving flexibility and efficiency at the same time and also does not give performance figures. [37] presents an ASIC design with a RISC CPU and cores for random number generation and modular arithmetic. Latter is word-based and may be used for RSA and ECC up to bit-widths of 2048, which promises easy upgradeability for new approaches like PBC. The design is aimed for client application making high throughput less important than low latency and this may be the reason why parallelization is considered to a lower degree than in this work.

3 Cryptographic Aspects

The public key cryptography schemes RSA and ECC are both based on modular arithmetic. ECC has the advantage that the used numbers are shorter, which results in a faster computation, especially for higher security levels. The downside of ECC is its more complex mathematical background, which makes it harder to understand and implement.

PBC builds on the foundations of ECC and extends it by the so called pairing, which is a way to compare pairs of points on elliptic curves, see [5]. An interesting application of PBC are identity-based schemes, which allow the public key to be derived from some unique identifier, e.g., an email-address. This would make the distribution of public keys unnecessary. The main reason PBC was chosen as third algorithm in this work is that it shares many parts with RSA and ECC, thus, allowing reusing most of the hardware cores and much of the software parts.

3.1 Abstractions Levels for Public Key Cryptography

The above introduced public key systems share some properties and may be classified into different abstraction levels as shown in Fig. 1. Note that RSA is the most simple approach, thus, does not need some parts required for ECC and PBC. In turn, PBC has additional elements compared to ECC.

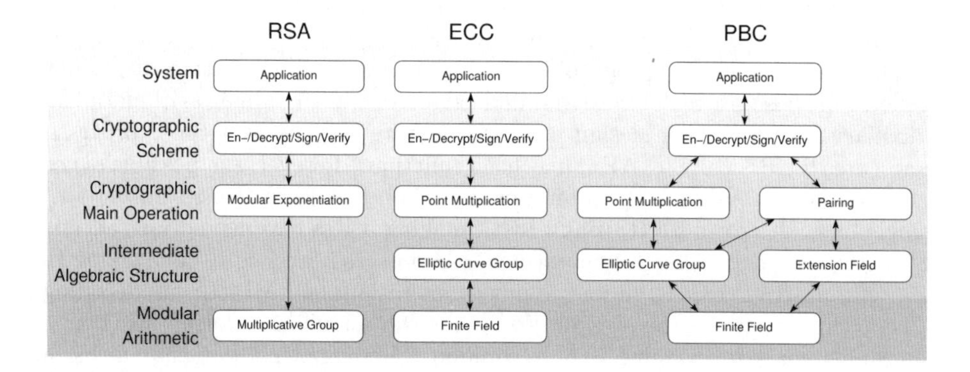

Fig. 1. Abstraction levels for different public key algorithms

The lowest abstraction level *modular arithmetic* contains modular operations, which are needed by algorithms on higher abstraction levels. For ECC and PBC the modular arithmetic must be a finite field, i.e., $\mathbb{GF}(q)$ in this work, where the modulus q is a prime number. For RSA the modulus n is not a prime, but the product of the large primes p and q. Thus, the modular arithmetic for RSA is not a finite field, but just a multiplicative group.

The abstraction level of the *intermediate algebraic structures* exists only for ECC and PBC. For former, this is the *elliptic curve group* offering operations on points on an elliptic curve. For PBC there are also operations in the *extension field*.

For RSA the abstraction level of the *cryptographic main operation* contains the *modular exponentiation*, which uses the modular arithmetic as a multiplicative group. The *point multiplication* used for ECC and PBC utilizes the elliptic curve group as an additive group. The *pairing* needed only for PBC uses operations of both the elliptic curve group and of the extension field.

On the next abstraction level complete *cryptographic schemes* are build by combining the cryptographic main operation(s) with cryptographic auxiliary functions like hash function and/or random number generation. Those are not shown in Fig. 1 for simplicity. On the *system* level, finally, the cryptographic schemes may be used by the application.

3.2 Parallelization in Public Key Cryptography

Performance improvements may be realized by utilizing parallelization. Consequently, many parallelization techniques for public key algorithms may be found in literature. In the following, several are presented categorized according to above abstraction levels. Note that parallelization may be exploited on different abstraction levels at the same time, as done in this work.

Modular Arithmetic Level. As modular multiplication is the most time consuming and most heavily used modular operation, a very obvious and common approach is to decrease execution time by applying many of its elementary building blocks – be they bit- or word-based – in parallel. Parallelization of the remaining modular operations is not as effective, as those are relatively simple and fast. Note that the following list does not constitute an exhaustive list.

Buses of Full Bit-Width. Basically, the complexity of a multiplication increases with the square of the utilized bit-length. If all bits are considered at the same time, the execution time is reduced to linear complexity. However, this requires also a proportional increase in terms of resources. For example, to double the bit-width of the implementation shown in [24] the amount of resources is also roughly doubled. Further examples for implementation with buses of full bit-width can be found in [28,3,23,26].

Parallelization using buses of full bit-width has relatively small costs, as it is done on a low level and most of the additional resources are really used to speed up computation and not for parallelizing the control logic as well. But with this approach the implementation is optimized for a single bit-width. Operations with larger bit-widths are difficult and for operations with a smaller bit-width, some of the resources stay unused. The architecture from [7] partly solves this problem as explained in Section 2. But it still exhibits some inflexibility as it does not solve the problem for larger bit-widths and leads to unused resources for bit-widths, which are not a multiple of the ECC-size.

Buses of full bit-width have another downside, if the memory structure has limitations, as it is the case on FPGAs. Although several memory blocks may be used to build up a memory with a wide enough data bus, the overall memory size is then likely much larger than required for cryptographic calculations. The alternative, namely realizing the memory in the logic cells, is more flexible regarding its widths and depths. But the used-up logic cells are then not available for other tasks.

For the computation of the pairing for PBC, the most commonly used algorithm is Miller's algorithm. Its approach is similar to that of the point multiplication adding some additional operations, see [5]. Because of that techniques allowing a parallelization of the point multiplication may also be used to parallelize Miller's algorithm.

Cryptographic Scheme and System Level. Parallelization on the cryptographic scheme level is suggested only seldom, because acceleration here has higher costs compared to lower levels, because of the higher portion of control logic. As described in Section 4 this work still utilizes parallelization on this level, because with the goals of flexibility and efficiency parallelization on lower levels may only be exploited to some degree. Contrary to lower levels, parallelization on cryptographic scheme level does not exhibit data dependencies, allowing as much parallelization as the resources allow.

Parallelization on system level is only mentioned because of completeness. It is obvious how the usage of a second whole system will speed-up the overall performance assuming the are no data dependencies.

4 Co-processor Architecture

The two main properties of the desired co-processor are high number of operations per second and high flexibility concerning the supported cryptographic schemes. The simultaneous support of RSA and ECC results in a wide range of possible bit-widths for the modular arithmetic. The security levels, which are reached with ECC with bit-lengths of 160 to 256 bit require bit-lengths of 1024 to 3072 bit for RSA. The co-processor architecture has to take this into account. Additionally, auxiliary functions like hash function and random number generation are needed to execute the complete cryptographic schemes inside a single chip.

4.1 Design Consideration

As high throughput is desired, the architecture should exploit all resources as continuously as possible, thus, minimizing the time some resources stay unused. The latency should still be decreased as long as this does not hinder throughput.

HW/SW Co-Design. Software implementations are relatively slow, but more flexible, because they are relatively easy to design and an increase in functionality does only increase the memory usage. In contrast, hardware is faster, but more inflexible, as it is more difficult to design and an increase in functionality requires additional state machines, of which only one is active at each point in time.

HW/SW co-design supports the combination of high throughput and high efficiency, because it allows to realize those parts in software and hardware, respectively, to which they are best suited to. The control-flow intensive tasks are implemented in software. The slowness of software is no drawback here, because the control-flow parts wait most of the time for the data-flow intensive parts to finish execution to start them anew with new operands. The high flexibility of software, in contrast, is an advantage, as the control-flow of the different cryptographic schemes differs significantly.

Consequently, the data-flow intensive parts are realized in hardware, as they are time-critical and very similar for the different schemes. Therefore, dedicated hardware is used to implement the modular arithmetic, which is the basis of all supported cryptographic schemes. The auxiliary functions are also realized in hardware: Although they are needed only occasionally, which would make them suited to a software implementation, they are also time-critical and may not be executed concurrently to other parts. Thus, while they would be executed in software, the hardware parts would stay unused, as the software processor is busy executing the auxiliary functions and cannot control the dedicated hardware.

The operations on the abstraction levels above the modular arithmetic are only control-flow intensive parts. They are different for the supported cryptographic schemes and most of the time wait for the execution of the modular operations. Consequently, they are implemented in software.

Parallelization. While the utilization of HW/SW co-design allows combining the goals flexibility and efficiency, parallelization may be used to increase the efficiency. The degree of parallelization should be chosen so that it aids the throughput of all supported cryptographic schemes. Generally, parallelization is less costly on lower abstraction levels, because less control-logic has to be paralellized as well. Therefore, starting from the modular arithmetic it is examined, how big the degree of parallelization can be.

On the lowest level a wide range of bit-widths must be supported. A parallelization degree optimized for the large bit-widths of RSA will lead to large amount of unused resources for ECC operations with their much smaller bit-widths. Therefore, the inflexible approaches like buses of full bid-width and pipelining of the inner loop will not aid throughput. However, pipelining of the outer loop allows to chose an arbitrary amount of pipeline stages, thus, relatively short pipelines optimized for the small bit-width are possible, which still speed-up computations with larger bit-widths.

On the abstraction level above modular arithmetic the usage of parallel multiplier instances leads to increases speed. As described above it is possible to exploit two parallel multipliers for RSA using the Montgomery Powering Ladder. In contrast, the ECC and PBC operations would be able to take advantage of up to three instances without resulting into a high amount of unused resources. But because RSA is not able to exploit a third multiplier, only two parallel modular multiplier instances are used.

ECC and PBC have one additional abstraction level compared to RSA. Here it would also be possible to use the Montgomery Powering Ladder (this time for point multiplication or the pairing) to exploit parallelization to gain an additional speed-up. But as the resources required for this would not be usable for RSA, no parallelization is utilized here.

The level of the cryptographic scheme is present for all three approaches and the degree of parallelization is not hindered by data dependencies. The only limitation is the amount of available resources and, therefore, the degree of parallelization should be chosen as high as those allow.

Summarizing, the concrete numbers of pipeline stages and parallel operating cryptographic schemes are still left open. As the optimal values for both are not obvious, possible combinations are examined in Section 6.

4.2 Resulting Architecture

Fig. 2 depicts the resulting architecture in general form. The time-critical modular operations are executed in ModArith cores, which feature two pipelined modules for modular multiplication and one module for the remaining modular operations. Each modular core is assigned to a Scheme Controller, which is realized by a general purpose processor. They execute the control-flow of the higher abstraction levels. The auxiliary cryptographic functions are implemented in the AES Core, which realizes symmetric encryption, hash function, and random number generation. Those operations are only a minor part of the complete cryptographic schemes and, therefore, the core may be shared by the Scheme

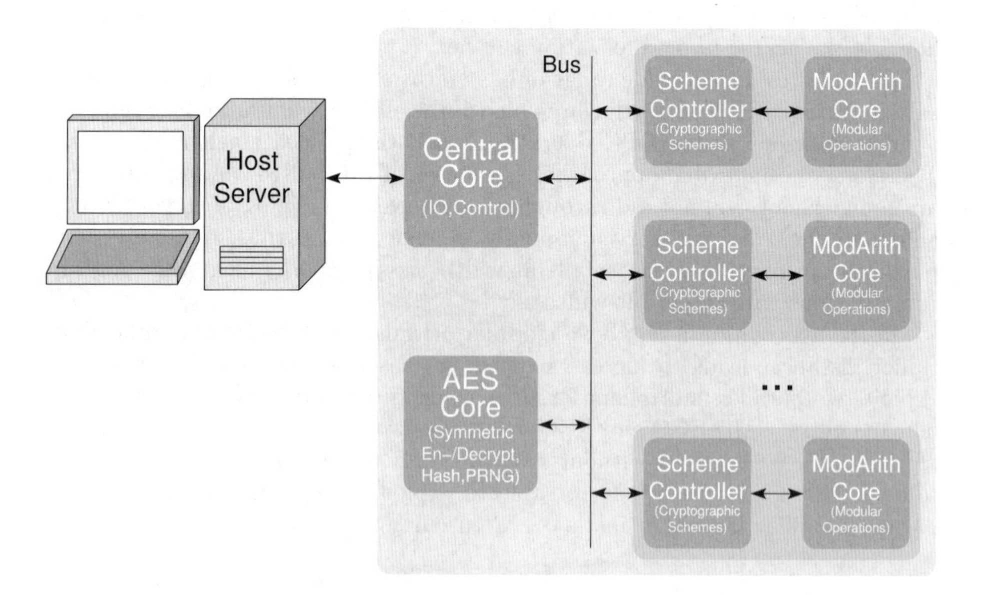

Fig. 2. Proposed co-processor architecture

Controllers. The Central Core, finally, controls the different Scheme Controllers and handles communication with the outside world.

Modular Arithmetic. The structure of the ModArith-cores is a refined version of that proposed in [16] and is depicted in Fig. 3. At its heart are the two modular multipliers ModMultA/B. Both are assigned their own memory allowing them unrestricted access via its first port as motivated in [16]. The second port of each memory is shared between the module for the Remaining Modular Arithmetic and the Scheme Controller. Latter one needs it to write and read the input values and the results, respectively, and may only access the memory, while the ModArith core is not executing.

Remaining Buildings Blocks. The core for the auxiliary cryptographic functions is taken from [17]. It supports symmetric en-/decryption, hash function, and random number generation. Its realization is aimed at minimized resources usage. This is achieved by using the AES block cipher as basis for all three functions, which allows them to share many resources.

The Scheme Controllers execute all operations above the modular arithmetic level and are realized with general purpose processors. Although each ModArith core is assigned to one Scheme Controller, it would also be possible to control all ModArith cores with a single Scheme Controller, as those wait for the completion

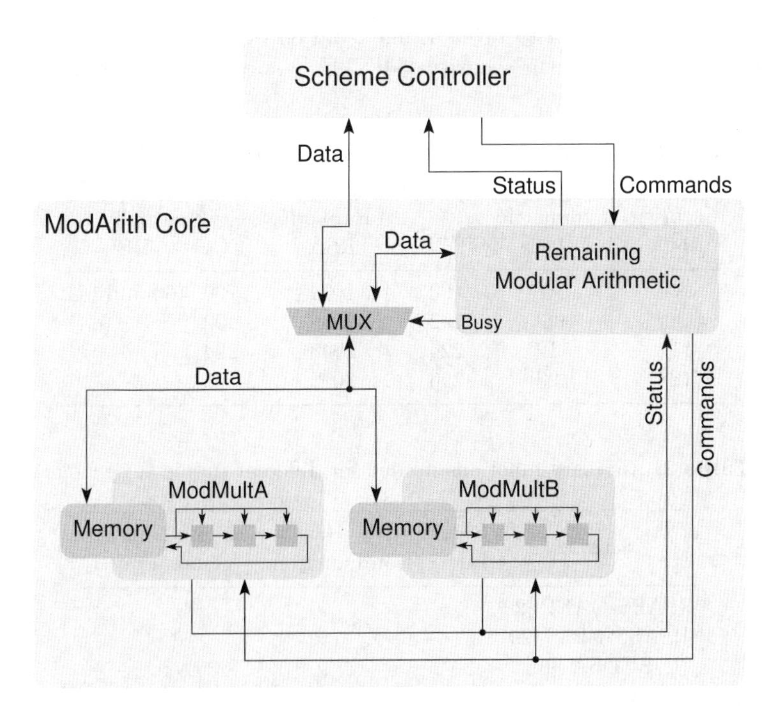

Fig. 3. ModArith core highlighting the memory architecture

of the modular operations most of the time anyway. It was decided against this, because then the software would have to support the concurrency of the hardware. The usual solution for this problem is the usage of threads. But the amount of cycles needed for a thread change is in the same order as the execution time of a modular operation. Thus, then the time-critical part, namely, the modular arithmetic would have to wait for the control-logic quite often.

The Central Core, which is responsible for communication with the outside world and for controlling all other cores, is also realized by a general purpose processor. Although those tasks could also be executed by the Scheme Controllers, it has three advantages to do this with a dedicated instance. It provides a fixed communication partner for the outside world, lets the Scheme Controllers concentrate on operating the ModArith cores as fast as possible, and the clear division of tasks eases implementation.

5 Prototype Implementation

To evaluate the proposed architecture, a prototype implementation was created, which is described in the following. It was realized using a XUP2P board [38] and the synthesis tools Xilinx EDK and ISE (both in version 9.1).

5.1 Supported Functionality

Usual recommended security levels range from 80 bit for medium term security up to 128 bit for long term security [34]. The respective key-length to achieve

Table 1. Different security bit-lengths and the resulting key-length

Security (bit)	Symmetric Cryptography	Hash	Public Key Cryptography RSA	ECC	PBC
80	80	160	1024	160	160 and 512
96	96	192	1536	192	–
112	112	224	2048	224	–
128	128	256	3072	256	–

Table 2. Auxiliary functions needed for different public key schemes

Scheme	Symmetric Encryption	Hash Function	Random Number Generation
RSAES-OAEP [29]	–	✓	✓
RSASSA-PSS [29]	–	✓	✓
ECDSA [9]	–	✓	✓
ECIES [10]	✓	✓	✓
BLS [6]	–	✓	✓

these security levels with the different cryptographic schemes are shown in Table 1. Thus, the modular operations have to be able to work with bit-widths between 160 and 3072 bit.

Some auxiliary functions are needed for the complete execution of the supported schemes. Table 2 depicts which are required for which scheme.

5.2 Implementation Overview

Fig. 4 gives an overview on the prototype implementation. It extends Fig. 2 by realization details like memory usage and type of the general purpose processor. The presented configuration utilizes four scheme controllers, as this is the maximum number, which was instantiated for the experiments.

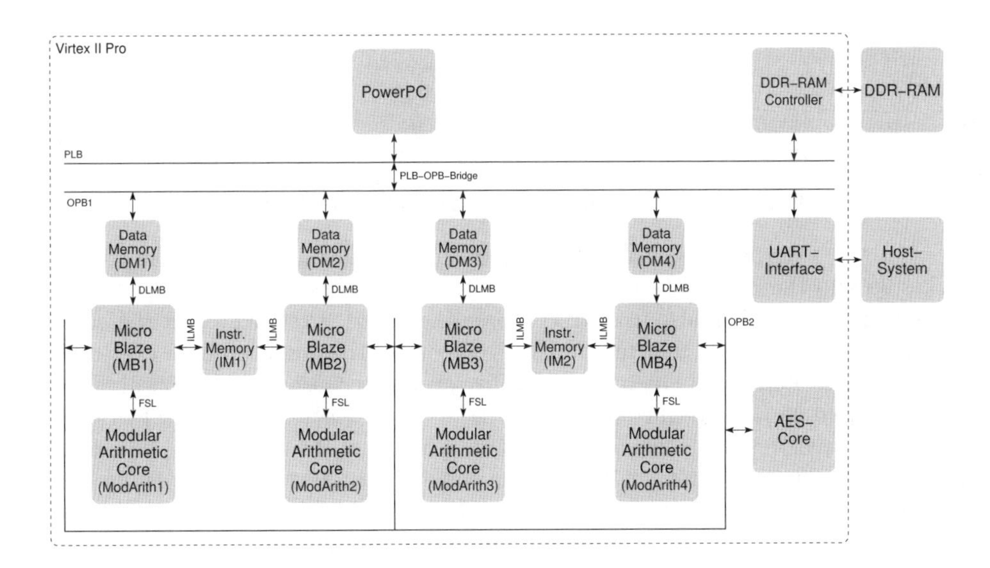

Fig. 4. Detailed view on the new co-processor architecture

The scheme controllers were realized using soft core processors called Micro Blaze. Their memories are implemented using the BRAM blocks connected to them via the Local Memory Bus (LMB). The ModArith cores and the AES core are realized as dedicated hardware in VHDL and connected using the Fast Simplex Link (FSL) and the Open Processor Bus (OPB), respectively. The central core, finally, uses a PowerPC core and uses off-chip memory. Note that this only makes sense for a proof-of-concept implementation, as it allows eavesdropping on the memory content of the central core. However, otherwise the on-chip memory would have only been sufficient to store the RSA and ECC functionality. Note also that a fully usable implementation should use a faster interface to the host server than an UART-interface.

5.3 Modular Multiplication

Because modular multiplication is the most critical operation for the supported cryptographic schemes, special care must be taken for its implementation. It is realized using the pipelining approach from [33], but works in a word-based manner to allow exploitation of the dedicated multipliers of the FPGA. Although such a realization can be found in [15], its memory access was not compatible to the memory architecture used for the prototype implementation. Therefore, a new implementation was created, which complies to the memory interface, see [27]. Note that this realization is not able to compute numbers with a word-count smaller than the double of the number of pipeline stages.

The pipeline of the modular multipliers with s stages can be seen in Fig. 5. The register *Reg* reduces the longest path, thus allows to operate the multiplier with a cycle frequency of 100 MHz.

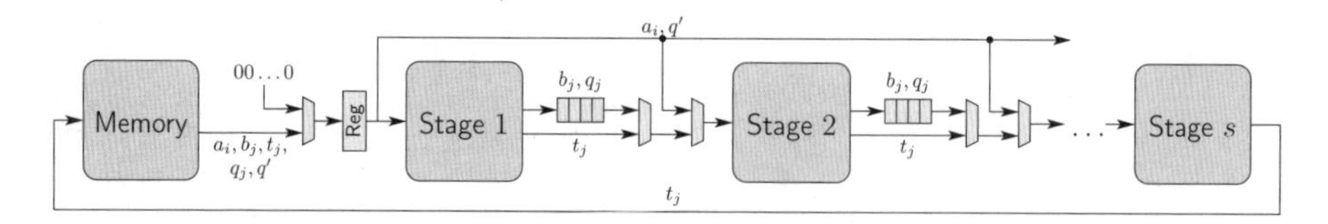

Fig. 5. Pipeline architecture of the modular multiplication

If $0 \leq k \leq s$, then the k-th stage computes the n inner steps of first the $(k-1)$-th step and then the $(k-1+s)$-th step of the outer loop. If $n > (k-1+s)$, the k-th stage is then also used to compute the $(k-1+2s)$-th step of the outer loop. If necessary, this continues until all n steps of the outer loop are executed.

Each j-th step of the inner loop takes five cycles, because five memory accesses are required for the intermediate result $t_j(i) := a_i \cdot b_j + u_i \cdot q_j + t_{j+1}(i-1) + carry$. Here $t_j(i)$ denotes the j-th word of t in the i-th step of the outer loop. Note that the stages after the first get their input values from their previous stage and could compute their result in only four cycles. But because this would lead to coordination problems for storing the results of the last stage in memory, it was decided that each stage takes the following five cycles for the computation of step j:

1. a_i is read from memory and stored in the stage requiring it.
2. b_j is read (from memory or from previous stage). It is used in $a_i \cdot b_j$ and also passed on to the next stage via the shift registers.
3. The intermediate result $t_{j+1}(i-1)$ is read (from memory or previous stage) and added to the intermediate result of this stage.
4. $t_j(i)$ computed by the last stage is written into memory.
5. q_j is read (from memory or previous stage) and used in $u_i \cdot q_j$ and also passed on to the next stage via the shift registers.

The shift registers buffer b_j and q_j for ten cycles until the following stage requires them, because the next stage needs b_j and q_j together with the intermediate result $t_{j+1}(i-1)$, which is calculated by the previous stage two steps of the inner loop later.

The datapaths of each stage are depicted in Fig. 6. If the path-widths do not fit, the values are filled with 0 from the left.

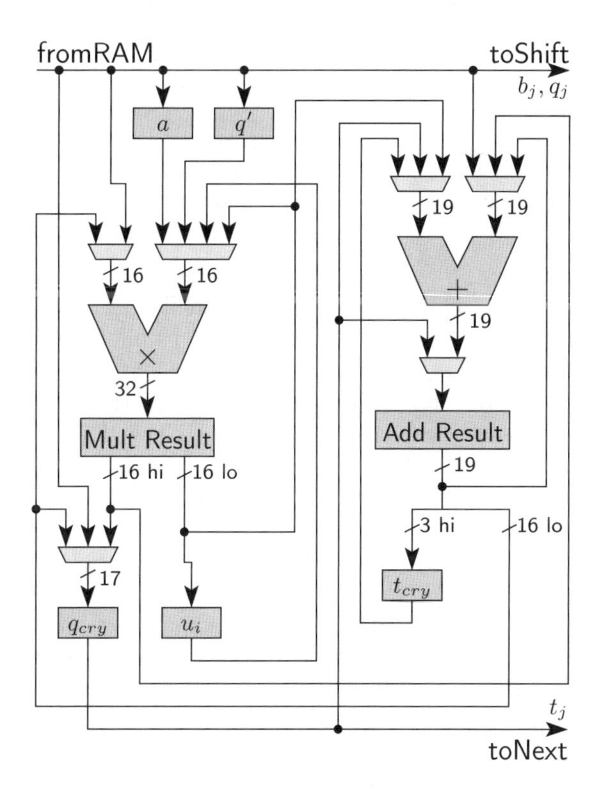

Fig. 6. Data path of a single pipeline stage

The total cycle count of a single modular multiplication is $5n \cdot \lceil \frac{n}{s} \rceil + 10\lceil \frac{n}{s} \rceil + 10s - 5$. Additionally, 3 to $5n - 2$ clock cycles are needed for the computation of the final correction of the Montgomery Multiplication (lines 6/7 of Algorithm 1).

6 Results

Table 3 shows the resource usage of four different configurations of the prototype implementation described above. In this context xSCyST denotes a design variant featuring x scheme controllers and y pipeline stages in each modular multiplier. The cycle frequency is 100 Mhz for all design variants. Table 3 also shows the resource usage of a single ModArith core.

Given an FPGA featuring more resources it would have been possible to instantiate variants with more scheme controllers and/or pipeline stages. For the variant with only 2 scheme controllers it is likely that more than 6 pipeline stages would have been synthesizable. But as the number of words must be at

Table 3. System resource usage

	Overall System		Single ModArith Core		
Variant Name	#Logic Cells	#BRAMs	#Logic Cells	#Word Multipliers	#BRAMs
4SC2ST	26344	117	3524	4	4
3SC3ST	24142	105	4406	6	4
3SC4ST	23666	105	5292	8	4
2SC6ST	21818	61	7168	12	4

least the double of the pipeline stages, it would have prohibited further bit-widths for ECC. With the utilized word-width of 16 bit ECC with 160 bit was already impossible.

Fig. 7 and Fig. 8 show the timing results of above variants for execution of signature and verification schemes, respectively. The values are averaged over 2000 scheme executions of each scheme. The throughput of the whole system is given as total operations per second, i.e., the number of operations all scheme controllers together execute in one second.

Variant 3SC4ST exhibits the best throughput for all schemes except for ECC with 160 bit, where 4SC2ST is better. Because 3SC4ST has as total of 24 pipeline stages while 4SC2ST has only 16, it seems that for short bit-length additional scheme controllers are more more effective than additional pipeline stages.

While 3SC4ST is the best variant for all scheme, the second-best for RSA and PBC is 2SC6ST, while 4SC2ST is the second-best for ECC. This is not surprising, because the large bit-widths of RSA and PBC are better able to take advantage of the long pipelines: Firstly, pipelined multipliers always compute a multiple of the number of stages in words, thus, the relation between the actually used bit-width and the required bit-width is smaller for larger bit-widths. Secondly, all stages of a pipeline work only concurrently after they are filled stage-by-stage. Therefore, computations with large bit-widths keep all stages filled longer.

Table 4 contains the resource and performance figures from variant 3SC4ST and from comparable designs from literature. The resources are given in logic cells (LC) for Xilinx FPGAs and logic elements (LE) for Altera FPGAs and those numbers are roughly comparable. Note that Table 4 only contains values for ECC and RSA, because designs for PBC concentrate on realizations over $\mathbb{GF}(2^m)$.

Most designs in literature are limited to the cryptographic main operation. From Table 4 only [13] gives values for complete RSA-encryption and decryption, which have a similar complexity as verification and signing, respectively. For the others only the timing for a point multiplication or a modular exponentiation is given, either with short or long exponent, which is used in verification or signing, respectively. The values marked with '*' are estimated, because the reference did only give the value for a single modular multiplication.

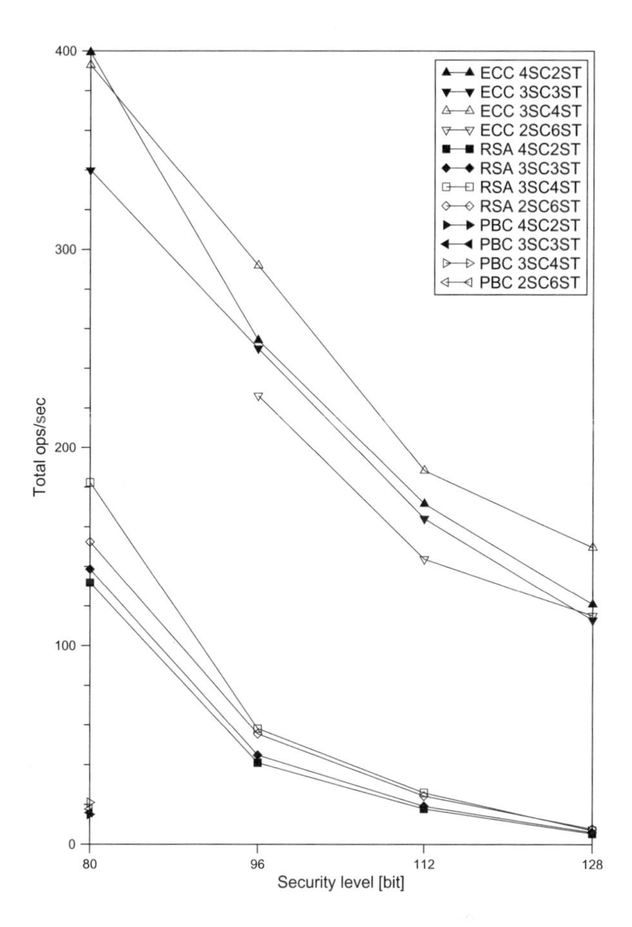

Fig. 7. Results for signing schemes

Contrary to the other references, [13] supports complete schemes by featuring cores for AES, SHA-1, RSA, and ECC over $\mathbb{GF}(2^m)$. Only the performance for RSA is comparable[1] and all variants from this work show better throughput for the RSA-signing operation, which is comparable to the RSA-decryption. The higher performance of [13] for the encryption (compared to RSA-verification) could stem from the utilization of a shorter exponent (only 5 bits in [13], 17 bits in this work).

Similar to this work, [7] also aims for hardware beneficial for both RSA and ECC at the same time. Again, all variants from this work show better throughput, probably because of the usage of the FPGA's dedicated word-multipliers.

[19,21] feature optimized designs for ECC and RSA, respectively, and have both a considerable higher throughput than 3SC4ST. But both are less flexible, as they are committed only to a single bit-width. Also, as they do concentrate on the cryptographic main operation, only the 15876 LC used for the modular

[1] Modular multiplications over $\mathbb{GF}(2^m)$ and $\mathbb{GF}(q)$ are considerably different.

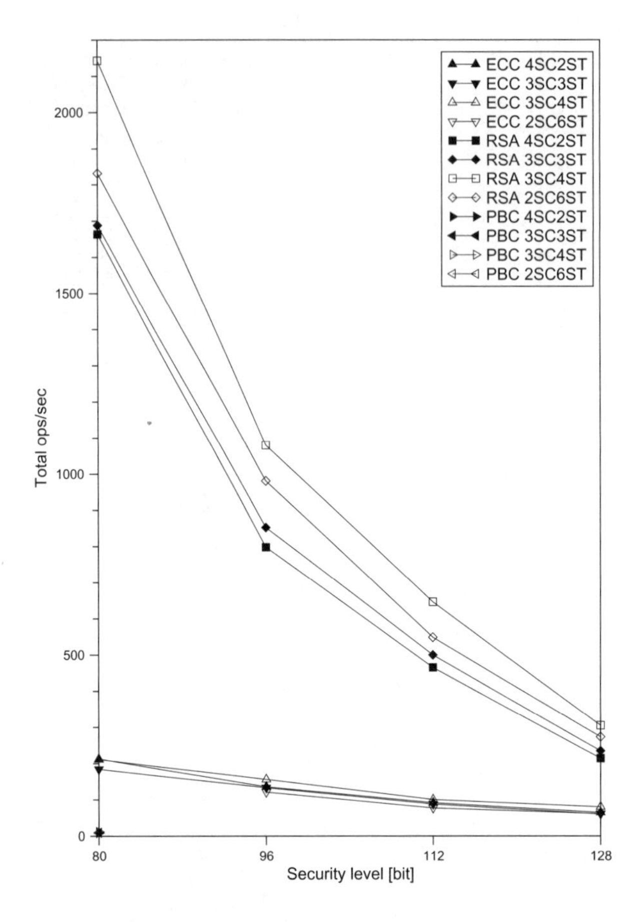

Fig. 8. Results for verification schemes

arithmetic in 3SC4ST should be counted in a comparison, making 3SC4ST being considerably smaller than both.

Similar to this work, [32] features two pipelined modular multipliers in parallel. But it considers only RSA. The performance figures shown in Table 4 are for a variant with 16 pipeline stages for each multiplier. Even if the lower resource usage of [32] is taken into account, 3SC4ST exhibits a higher throughput.

A pipelined modular multiplier able to work on values of arbitrary bit-width, is presented in [12]. The values shown in Table 4 are those of a variant with 16 pipeline stages. It has a considerably higher throughput than 3SC4ST, if its lower resource usage is taken into account. This is probably because [12] exploits both more word-multipliers and more memory blocks, which were very scarce in the prototype implementation of this work, however.

Table 4. Performance figures of comparable designs

Ref	LC/LE	Bit-width	Operation Type	Latency	Ops/Sec
this work	23666 LC	1024	ModExp (short Exp)	1.11 ms	2696.63
			ModExp (long Exp)	16.18 ms	185.40
		2048	ModExp (short Exp)	4.12 ms	727.89
			ModExp (long Exp)	115.32 ms	26.02
		256	PointMult	17.43 ms	172.16
[13]	28451 LE	1024	RSA-Encr	0.25 ms	4000.00
			RSA-Decr	31.93 ms	31.32
[7]	10534 LC	256	PointMult	*26.50 ms	*39.06
		1033	ModExp (long Exp)	*35.76 ms	*27.96
[19]	31510 LC	256	PointMult	3.86 ms	259.07
[21]	24324 LC	1024	ModExp (long Exp)	3.829 ms	261.16
[32]	4112 LE	1024	ModExp (short Exp)	7.8 ms	128.21
			ModExp (long Exp)	39 ms	25.64
		2048	ModExp (short Exp)	31 ms	32.26
			ModExp (long Exp)	222 ms	4.50
[12]	4768 LC	1024	ModExp (long Exp)	8.17 ms	122.40

7 Conclusion

This work proposed a novel co-processor architecture aimed for public key cryptography and server application. Contrary to the usually approach in literature, it tries to achieve efficiency and flexibility at the same time. This limits the parallelization degree on lower abstraction levels, which is compensated by also utilizing parallelization on higher abstraction levels. The validity of this approach is shown by means of an SoC-based prototype implementation able to execute complete cryptographic schemes on-chip and which is only outperformed by highly specialized realizations geared for a single bit-width, thus, lacking a comparable flexibility.

References

1. Aoki, K., Hoshino, F., Kobayashi, T., Oguro, H.: Elliptic Curve Arithmetic Using SIMD. In: Davida, G.I., Frankel, Y. (eds.) ISC 2001. LNCS, vol. 2200, pp. 235–247. Springer, Heidelberg (2001)
2. Batina, L., Bruin-Muurling, G., Örs, S.B.: Flexible Hardware Design for RSA and Elliptic Curve Cryptosystems. In: Okamoto, T. (ed.) CT-RSA 2004. LNCS, vol. 2964, pp. 250–263. Springer, Heidelberg (2004)
3. Bednara, M., Daldrup, M., von zur Gathen, J., Shokrollahi, J., Teich, J.: Reconfigurable Implementation of Elliptic Curve Crypto Algorithms. In: International Parallel and Distributed Processing Symposium (IPDPS) (2002), `http://www-math.upb.de/~aggathen/Publications/raw02.pdf`

4. Blake, I.F., Seroussi, G., Smart, N.P.: Elliptic Curves in Cryptography. Cambridge University Press, New York (1999)
5. Blake, I.F., Seroussi, G., Smart, N.P.: Advances in Elliptic Curve Cryptography. Cambridge University Press, New York (2005)
6. Boneh, D., Lynn, B., Shacham, H.: Short Signatures from the Weil Pairing. In: Boyd, C. (ed.) ASIACRYPT 2001. LNCS, vol. 2248, pp. 514–532. Springer, Heidelberg (2001)
7. Crowe, F., Daly, A., Marnane, W.P.: A Scalable Dual Mode Arithmetic Unit for Public Key Cryptosystems. In: International Conference on Information Technology: Coding and Computing (ITCC), vol. 1, pp. 568–573 (2005)
8. Dutta, R., Barua, R., Sarkar, P.: Pairing-Based Cryptographic Protocols: A Survey, IACR eprint archive, 2004/064 (2004), http://eprint.iacr.org/2004/064
9. IEEE: IEEE 1363-2000 - Standard Specifications for Public-Key Cryptography, New York, USA (2000), http://grouper.ieee.org/groups/1363/
10. IEEE: IEEE 1363a-2004: Standard Specifications for Public-Key Cryptography – Amendment 1: Additional Techniques, New York, USA (2004), http://grouper.ieee.org/groups/1363/
11. Fischer, W., Giraud, C., Knudsen, E.W.: Parallel scalar multiplication on general elliptic curves over $\mathbb{F}_p$ hedged against Non-Differential Side-Channel Attacks, IACR eprint archive 2002/007 (2002), http://eprint.iacr.org/2002/007.pdf
12. Güdü, T.: A new Scalable Hardware Architecture for RSA Algorithm. In: International Conference on Field Programmable Logic and Applications (FPL), pp. 670–674 (2007)
13. Hani, M.K., Wen, H.Y., Paniandi, A.: Design and Implementation of a Private and Public Key Crypto Processor for Next-Generation IT Security Applications. Malaysian Journal of Computer Science 19(1), 20–45 (2006)
14. Joye, M., Yen, S.-M.: The Montgomery Powering Ladder, Workshop on Cryptographic Hardware and Embedded Systems (CHES). In: Kaliski Jr., B.S., Koç, Ç.K., Paar, C. (eds.) CHES 2002. LNCS, vol. 2523, pp. 291–302. Springer, Heidelberg (2003), http://www.gemplus.com/smart/rd/publications/pdf/JY03mont.pdf
15. Kelley, K., Harris, D.: Very High Radix Scalable Montgomery Multipliers. In: International Workshop on System-on-Chip for Real-Time Applications (IWSOC), Washington, DC, USA, pp. 400–404 (2005)
16. Laue, R., Huss, S.A.: Parallel Memory Architecture for Elliptic Curve Cryptography over $\mathbb{GF}(p)$ Aimed at Efficient FPGA Implementation. Journal of Signal Processing Systems 51(1), 39–55 (2008)
17. Laue, R., Kelm, O., Schipp, S., Shoufan, A., Huss, S.A.: Compact AES-based Architecture for Symmetric Encryption, Hash Function, and Random Number Generation. In: International Conference on Field Programmable Logic and Applications (FPL), Amsterdam, Netherlands, pp. 480–484 (2007)
18. Lim, C.H., Lee, P.J.: More Flexible Exponentiation with Precomputation. In: Desmedt, Y.G. (ed.) CRYPTO 1994. LNCS, vol. 839, pp. 95–107. Springer, Heidelberg (1994), citeseer.ist.psu.edu/lim94more.html
19. McIvor, C.J., McLoone, M., McCanny, J.V.: Hardware Elliptic Curve Cryptographic Processor Over $\mathbb{GF}(p)$. IEEE Transactions on Circuits and Systems I: Regular Papers 53(9), 1946–1957 (2006)
20. Menezes, A.J., van Oorschot, P.C., Vanstone, S.A.: Handbook of Applied Cryptography. CRC Press, Boca Raton (1997), http://www.cacr.math.uwaterloo.ca/hac/

21. Michalski, A., Buell, D.: A Scalable Architecture for RSA Cryptography on Large FPGAs. In: IEEE Symposium on Field-Programmable Custom Computing Machines (FCCM), pp. 331–332 (2006)
22. de Macedo Mourelle, L., Nedjah, N.: Efficient Cryptographic Hardware Using the Co-Design Methodology. In: International Conference on Information Technology: Coding and Computing (ITCC), vol. 2, pp. 508–512 (2004)
23. Nedjah, N., de Macedo Mourelle, L.: Fast Less Recursive Hardware for Large Number Multiplication Using Karatsuba-Ofman's Algorithm. In: Yazıcı, A., Şener, C. (eds.) ISCIS 2003. LNCS, vol. 2869, pp. 43–50. Springer, Heidelberg (2003)
24. Örs, S.B., Batina, L., Preneel, B., Vandewalle, J.: Hardware Implementation of a Montgomery Modular Multiplier in a Systolic Array. In: International Parallel and Distributed Processing Symposium (IPDPS), p. 184 (2003), http://www.cosic.esat.kuleuven.be/publications/article-32.pdf
25. Ohba, N., Takano, K.: An SoC design methodology using FPGAs and embedded microprocessors. In: Conference on Design automation (DAC), pp. 747–752 (2004)
26. Orlando, G., Paar, C.: A Scalable $\mathbb{GF}(p)$ Elliptic Curve Processor Architecture for Programmable Hardware. In: Koç, Ç.K., Naccache, D., Paar, C. (eds.) CHES 2001. LNCS, vol. 2162, pp. 348–363. Springer, Heidelberg (2001)
27. Rieder, F.: Modular Multiplikation mit kurzen Pipelines, Technische Universität Darmstadt, Diplom Thesis (2008) (in German), http://www.vlsi.informatik.tu-darmstadt.de/staff/laue/arbeiten/rieder_thesis.pdf
28. Rodríguez-Henríquez, F., Koç, Ç.K.: On fully parallel Karatsuba Multipliers for $GF(2^m)$. In: International Conference on Computer Science and Technology (CST), pp. 405–410 (2003), http://security.ece.orst.edu/papers/c29fpkar.pdf
29. RSA Laboratories: PKCS #1 v2.1: RSA Cryptography Standard (June 2002), ftp://ftp.rsasecurity.com/pub/pkcs/pkcs-1/pkcs-1v2-1.pdf
30. Saqib, N.A., Rodríguez-Henríquez, F., Díaz-Pérez, A.: A Parallel Architecture for Computing Scalar Multiplication on Hessian Elliptic Curves. In: International Conference on Information Technology: Coding and Computing (ITCC), vol. 2, pp. 493–497 (2004)
31. Sakiyama, K., Batina, L., Preneel, B., Verbauwhede, I.: HW/SW Co-design for Accelerating Public-Key Cryptosystems over GF(p) on the 8051 μ-controller. In: Proceedings of World Automation Congress (WAC) (2006)
32. Šimka, M., Fischer, V., Drutarovský, M.: Hardware-Software Codesign in Embedded Asymmetric Cryptography Application – a Case Study. In: Y. K. Cheung, P., Constantinides, G.A. (eds.) FPL 2003. LNCS, vol. 2778, pp. 1075–1078. Springer, Heidelberg (2003), http://citeseer.ist.psu.edu/simka03hardwaresoftware.html
33. Tenca, A.F., Koç, Ç.K.: A Scalable Architecture for Modular Multiplication Based on Montgomery's Algorithm. IEEE Trans. Computers 52(9), 1215–1221 (2003), http://security.ece.orst.edu/papers/c17asamm.pdf
34. Vanstone, S.A.: Next generation security for wireless: elliptic curve cryptography. Computers & Security 22(5), 412–415 (2003)
35. Walter, C.D.: Improved linear systolic array for fast modular exponentiation. IEE Proceedings Computers & Digital Techniques 147(5), 323–328 (2000)
36. Wannemacher, M.: Das FPGA-Kochbuch. MITP-Verlag (1998)
37. Wu, M., Zeng, X., Han, J., Wu, Y., Fan, Y.: A High-Performance Platform-Based SoC for Information Security. In: Conference on Asia South Pacific design automation (ASP-DAC), pp. 122–123 (2006)
38. XILINX: Xilinx XUP Virtex-II Pro Development System, http://www.xilinx.com/univ/xupv2p.html

Cellular-Array Implementations of Bio-inspired Self-healing Systems: State of the Art and Future Perspectives

André Seffrin[1] and Alexander Biedermann[2]

[1] CASED (Center for Advanced Security Research Darmstadt)
Mornewegstraße 32
64289 Darmstadt, Germany
andre.seffrin@cased.de

[2] Technische Universität Darmstadt
Department of Computer Science
Integrated Circuits and Systems Lab
Hochschulstraße 10
64289 Darmstadt, Germany
biedermann@iss.tu-darmstadt.de

Abstract. This survey aims to give an overview of bio-inspired systems which employ cellular arrays in order to achieve redundancy and self-healing capabilities. In spite of numerous publications in this particular field, only a few fundamentally different architectures exist. After a general introduction to research concerning bio-inspired systems, we describe these fundamental system types and evaluate their advantages and disadvantages. In addition, we identify areas of interest for future research.

1 Introduction

In various technical fields, mechanisms inspired by biology have been employed. As these mechanisms have been successful in the natural world, it seems logical to assume that they will also be relevant when applied to systems conceived by man. The most outstanding features of biological systems are their capability of self-healing, self-replication and adaptability. Self-healing systems which rely on traditional methods, such as TMR (*triple modular redundancy*) [9], usually have a high resource overhead. In many bio-inspired systems, the overhead can be fine-tuned much better according to safety needs than within regular systems. In this survey, we focus on architectures of cellular arrays which provide redundancy through provision of spare cells. The prime institutes to conduct research in this field include the University of York, England and the Swiss Federal Institute of Technology.

While a large number of publications exist in this field, the underlying mechanisms are usually quite similar. Unfortunately, it must be noted that within a lot of papers on this topic, a huge amount of redundancy exists. In addition, the biological terminology used for the description of certain hardware mechanisms are not entirely consistent.

A. Biedermann and H. Gregor Molter (Eds.): Secure Embedded Systems, LNEE 78, pp. 151–170.
springerlink.com

This survey is structured as follows: We present different classification structures for bio-inspired systems in section 2, and give an overview of hardware-software analogies in section 3. In section 5, we examine different types of self-healing arrays and evaluate them with regard to features such as redundancy and self-healing mechanisms. Design methodologies, a topic still bearing many opportunities for cellular-array type structures, are considered in section 8. A conclusion and future outlook are given in section 9.

2 Classification of Bio-inspired and Self-healing Systems

The most well-suited classification system for the type of system to be considered in this survey stems from Sanchez et al. [32]. Three fundamental properties of biological systems, which may aid as inspiration for hardware systems, are identified: *Phylogeny*, *ontogeny*, and *epigenesis*. Phylogeny refers to the fact that the genetic code of species can evolve over subsequent generations. This process is driven by the three mechanisms of *selective reproduction*, *crossover* and *mutations* [41]. Ontogeny refers to the growth processes of a single cell, in particular through the methods of *cellular division* and *cellular differentiation*. Finally, epigenesis implies that an organism can evolve through learning, which is usually accomplished by neural networks.

These three properties can be visualized along three coordinate axes, resulting in Figure 1 (as first presented in [32]). This method of system classification is referred to as the POE model. The majority of the systems described within this survey lie solely on the ontogenetic axis and are indicated within the coordinate system.

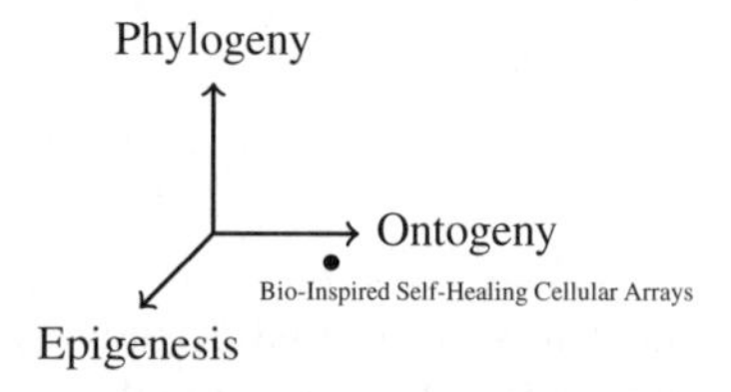

Fig. 1. Phylogenetic, Ontogenetic and Epigenetic Axes

For a an overview on systems focusing on the phylogenetic axis, see the survey by Yao et al. [42]. For a survey which mainly explores the epigenetic axis of bio-inspired hardware, see [13]. Typical high-level means for performing the classification of self-healing systems [20] are tailored to high-level software systems and not applicable to cellular-array based designs at their current level of sophistication.

Bio-inspired cellular arrays are organized in a homogeneous structure typically called *cellular lattice* [12], a term which has no direct biological counterpart, since cells in nature are not arranged in regular, grid-like structures. With reference to the POE-model, this type of versatile, cellular array has also been called *poetic tissue* [41] and *cell matrix architecture* [16].

3 Analogies between Biology and Hardware Systems

Various mechanisms from nature are proposed for use in self-healing cellular arrays. The term *embryonics*, introduced by Mange et al. [17, 18, 19], describes the construction of systems by using homogeneous cells. Like natural cells, these cells contain identical DNA, i.e., operating instructions, and differentiate their behaviour according to their position within the system. The method of having each cell in the homogeneous array carry the same "genetic code" is widely used in cellular arrays, and the cell configurations are usually referred to as DNA or *operative genome* [38]. The cells themselves, capable of acquiring very different types of functionality, can be considered to be *stem cells* [12, 16, 30]. The ability of cells to self-replicate can be compared to the biological feature of *mitosis* (DNA replication) [26].

The following analogy with regard to self-healing mechanisms is presented in [14, 15]: In the immune systems of vertebrates, pathogens, which can be compared to hardware errors in computing systems, are detected by lymphocytes known as *B-cells*. After the pathogens have been marked, they can be destroyed by another type of lymphocyte called *T-cell*. In hardware, this mechanism may be represented by a replicated module which emulates the behaviour of a module under supervision. Such a module could detect behaviour deviations and thus detect foreign cells. For the sake of completeness, we state that lymphocytes are a subspecies of white blood cells which possess various methods of dealing with pathogens. No further mechanisms with regard to white blood cells are mentioned within the literature that is the basis of this survey.

A different principle is mentioned in [11, 12], which proposes the analogy of endocrine communication and paracrine signalling. The endocrine system works via glands which produce hormones. The hormones are then distributed via the bloodflow throughout the body. Special receptors at target cells receive the hormones and trigger the corresponding responses. While endocrine communication works over long distances, paracrine signalling is restricted to very short distances. In the case of paracrine signalling, the messaging molecules do not enter the bloodstream. In hardware, endocrine communication may be modelled by a data packet with a distinct marker which travels through an on-chip network. Paracrine signalling may be implemented in hardware by a simple point-to-point communication over a wire.

A comparison of the different levels of hierarchy at the biological and at the hardware level is given in figure 2, originally presented in [24]. The term *cicatrization* refers to self-repair of cells at the molecular level [40].

Further analogies between biological systems and technical means of self-healing can be found in a survey by Ghosh et. al [10], which focuses on self-healing on the software level.

4 Error Detection and Correction

Error correction within a bio-inspired system is usually executed at the level of cellular granularity, i.e., each cell is responsible to detect errors within itself. The following methods are traditionally used for error correction and detection in security-critical systems:

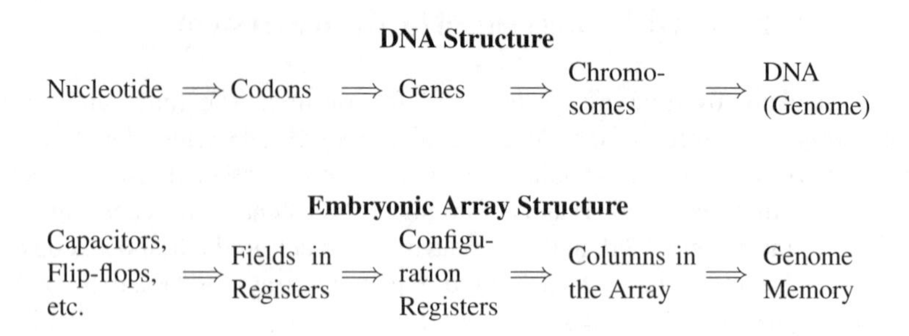

Fig. 2. Comparison of Terms in Biology and Hardware Design

- **Built-In Self Test:** The term BIST *(Built-In Self Test)* commonly refers to methods of self testing which use test vectors for system testing. This means that it is known how a system will respond to certain input combinations. For small systems such as those employed in bio-inspired cells with a low level of granularity, full system coverage can be achieved. BIST systems have the drawback that testing at run-time is not easy to implement.
- **Parity Calculation:** The system configuration of each cell is stored within registers. They are prone to radiation-induced faults which manifest themselves as bitflips. By calculating the parity value of the stored configuration, it can be determined whether it is still valid. For a bio-inspired cell, it may be practical to simply perform an *xor* on all included bits, or to employ a CRC check. For large-scale systems with extended configuration files, various crypthographic hash mechanisms may be applied. This method requires a certain amount of data redundancy, as it also requires the parity data to be stored within the cells.
- **Module Redundancy:** A popular way of providing security by redundancy is to instantiate multiple hardware modules of the same type, and to use a majority voter to decide which output to use. The most common variant is called TMR (*triple modular redundancy*), and is illustrated in figure 3. In a reduced variant of this scheme, only two copies of the same design are present, and their outputs are compared for equivalence. This type of system can only detect, but not correct errors. This approach is commonly known as DWC (*duplication with comparison*).

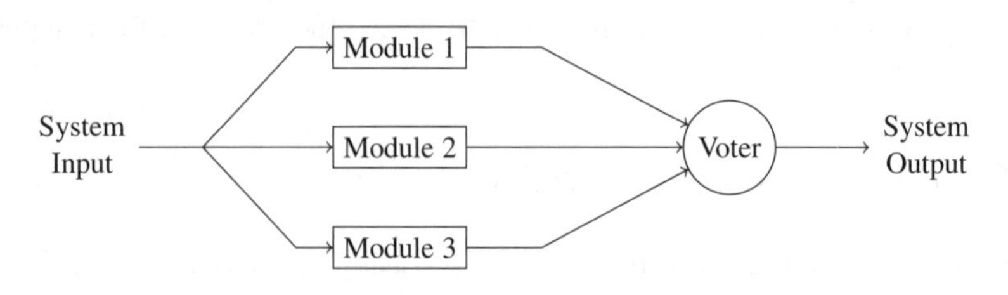

Fig. 3. Triple Modular Redundancy

Of the approaches listed above, the most common one to be applied in bio-inspired systems is DWC. A variant of BIST methods is that of *self-nonself discrimination* [8]. This method assumes that systems have the two fundamental sets of states *self* and *non-self*. When a healthy system is probed by a set of test-vectors, the corresponding healthy states can be determined. In biological systems, the protective agents are exposed to the body's molecules (i.e., the *self* part) during their gestation period, and agents undergo a period of *negativ selection* in order to remove obtain a set of agents which only tolerates the valid system configuration [7].

This method of error detection has been applied to electronics under the term of *immunotronics* [2, 3]. Its application in conjunction with cellular arrays has only been demonstrated in [4,6], although it must be noted that this method is merely employed as a security layer above the cellular array, and does not serve to protect individual cells.

5 Architectures for Redundant, Cellular Arrays

The standard design for embryonic cells is a matrix of X by Y cells. We also present two other types of designs which function in a slightly different way. These designs impose a strong locality on cells. Finally, we represent a higher-level architecture which utilizes the endocrine and paracrine messaging. We give formulas for calculating the degree of redundancy and hardware overhead for each of the low-level designs.

5.1 MUXTREE Architecture

The MUXTREE architecture [18, 29] is a type of bio-inspired self-healing cellular array which has been quite thoroughly researched. In this type of design, cells are laid out in a homogeneous grid, and cell replacement is implemented by shifting all cells within a row or column. Depending on the architecture type, a shift by one or multiple cells is possible. Some architectures offer shifts in both directions. Each cell carries all the different data sets for cell configurations that it may have to adopt. Thus, resource requirements in this architecture decrease with higher locality of redundancy, and also because fewer routing resources are required. For instance, a system may only allow up to n cell shifts within each row.

Within this type of cellular array, it is possible to identify two prime cell eliminations schemes, *cell elimination* and *row elimination*. Row elimination assumes that a complete row is replaced by a neighbouring row in case of a cell error. This is possible because the system provides a certain amount of spare rows. In most implementations, damaged cells become *transparent*, i.e., they simply route data through themselves. This saves resources in contrast to the usage of separate channels capable of routing data around damaged cells. It is assumed that routing is rather failure tolerant in comparison to the remaining cell components. Thus, no tests with regard to proper routing are performed by any architecture under review in this survey. The method of row elimination is illustrated in figure 4. Cell elimination assumes that each row can tolerate a fix amount of cell faults, as each row contains a number of spare cells. This elimination method is illustrated in Figure 5.

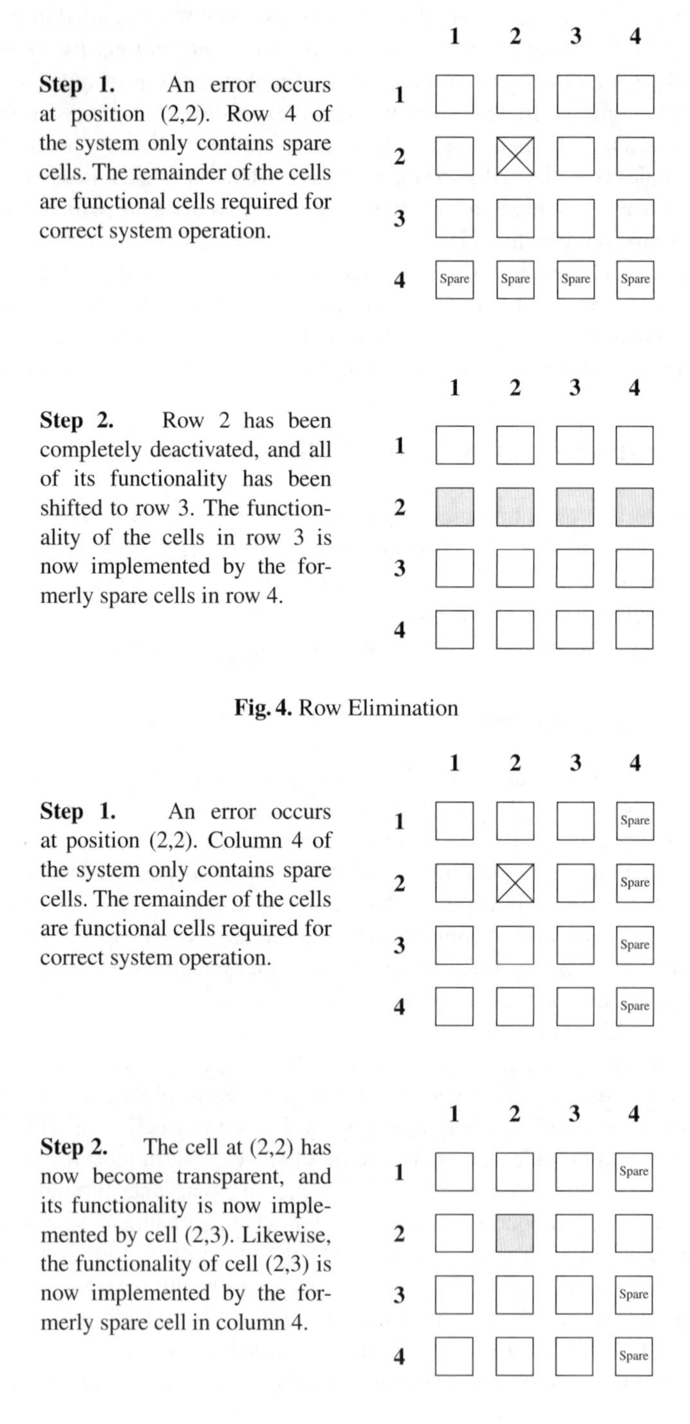

Fig. 4. Row Elimination

Fig. 5. Cell Elimination

For these schemes, various performance measurements can be given. The derivations for these formulas are presented in [25, 28]. Consider a square homogeneous array of cells with a dimension of n. For the case that x rows are eliminated, the percentage of cells lost can be calculated as follows:

$$F(n,x) = \frac{x \times n}{n^2} \times 100 = \frac{x}{n} \times 100 \tag{1}$$

$F(n,x)$ asymptotically decreases with increasing array sizes and thus, for sufficiently large arrays, the relevance of x diminishes. For cell elimination mode, no losses occur except for the damaged cells.

For a system which eliminates a row in case of an error occuring therein, let r be the number of rows required for the system to function properly, and n the total number of rows in the array. The amount of elements in each row is specified by m, and each of those elements possesses the failure rate λ_R. With these values, the system reliability can be calculated using the following formula:

$$R_{tr}(t) = \sum_{j=r}^{n} \binom{n}{j} e^{-jm\lambda_R t}(1 - e^{-m\lambda_R t})^{n-j} \tag{2}$$

In contrast to the row-elimination scheme, within the cell-elimination scheme, a row is only eliminated when all individual cells within a row have been corrupted. Formula 3 gives the reliability of such a system with respect to time, λ_C denotes the failure rate of every cell which performs array elimination. It requires formula 4 for the calculation of R_{rc}, the reliability of each individual row.

$$R_{tc}(t) = \sum_{j=r} \binom{n}{j} R_{rc}(t)^j (1 - R_{rc}(t))^{n-j} \tag{3}$$

$$R_{rc}(t) = \sum_{i=k}^{m} \binom{m}{i} e^{-i\lambda_C t}(1 - e^{\lambda_C t})^{m-i} \tag{4}$$

Cell elimination yields a much higher reliability than row elimination as only a faulty cell deactivated instead of a complete row containing still functioning cells. However, the routing and logic overhead of cell elimination is high compared to row elimination, so that only few systems rely on this method. In [43], a system is presented which supports both row and cell elimination, i.e., after a vital amount of cells within one row has been eliminated, the row itself is eliminated and replaced by a neighbouring row.

The redundancy overhead introduced by this type of cellular architecture is highly variable, and easy to control. In order to introduce redundancy to a system with $X \times Y$ squares, we can add R_c spare columns, with a resulting overhead of $R_c Y/XY$. If we want to ensure that complete rows can be replaced once all their redundant cells have been used up, we can add R_r rows, resulting in an overhead of $(R_r(X + R_c) + R_c Y)/XY$.

5.2 Architecture due to Szasz et al.

We now consider the design proposed in [35, 34, 36, 37], which features cells that are arranged in a grid structure. Each cell has a design which is illustrated in Figure 6. This arrangement is called *macro group*.

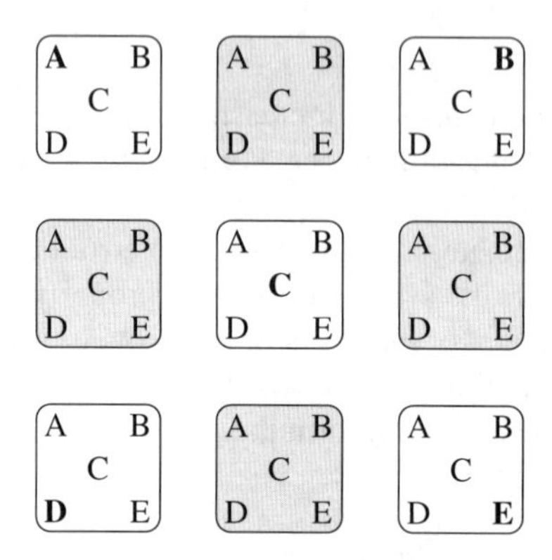

Fig. 6. Architecture due to Szasz et al.

Each cell with a white background is an active cell that contains five different genes. According to its position, one of these five genes is active, determining the functionality of the cell (active gene is bold). Besides the five active cells, there are 4 backup cells. Whenever any of the standard cells gets damaged, the backup cells can take over the functionality of the damaged cells.

This architecture provides a high amount of backup cells at a local level. While the cells in the previously examined architecture are relatively complex, each containing the full system gene or at least the genes required for the current row or column, these cells only store five local genes. Thus, cell size is reduced, but the degree of redundancy is increased. Calculating the overhead as the ratio of spare cells to functional cells, we obtain $4/5$.

5.3 Architecture due to Lala et al.

This architecture has been presented in [14, 15]. It combines local redundancy without the use of specific macro groups, and is illustrated in Figure 7 (connections between cells are not pictured).

Each functional cell (F) has two spare cells (S) and two router cells (R) as its direct neighbours in the horizontal and vertical directions. Each spare cell can serve to replace one of four functional cells which are located right next to it. Each functional cell only carries its own genetic information, while each spare cell carries the genetic information for all four cells whose functionality it may have to replace. The router cells ensure that the proper interconnections are retained in case of cell replacement.

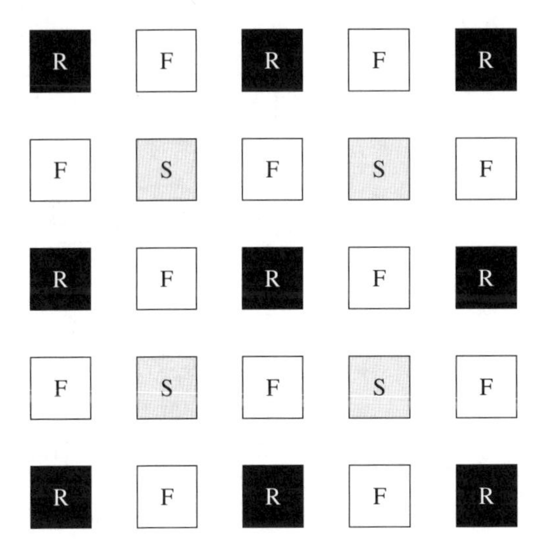

Fig. 7. Architecture due to Lala et al.

The overhead for this cell architecture is given as the ratio of spare cells to the sum of functional and router cells, which amounts to $1/3$. However, it should be considered that the elements included in router cells are not counted as area overhead in other architectures, in which routing elements are implemented outside of cells.

5.4 Architecture due to Greensted et al.

There are multiple other self-healing architectures which deviate from the general cell-scheme as introduced above, or which are especially adapted for their target platforms. One notable architecture is the design introduced in [12], which employs endocrine and paracrine messaging, as described in section 2. Each node in this network is pluripotent, and nodes are grouped into functional groups called *organs*. Each bionode has connections to its nearest eight neighbours. Figure 8 illustrates the architecture. Any organ can use the pluripotent cells as backup in case of cell failure.

Communication between organs is achieved by paracrine messaging, i.e., network packets are sent among the nodes. These paracrine messages are routed according to weighted random decisions, introducing an artificial blood flow. Since they are not supposed to remain within the blood flow indefinitely, the nodes will stop propagating after a certain amount of hops has elapsed. Paracrine is used for communication within organs: These messages are only propagated by one hop, but they are propagated to all neighbours of a cell. Due to the complex nature of this architecture, it does not lend itself to very fine-granular approaches. However, the paracrine system seems to be a viable solution for the distribution of information within large cellular arrays.

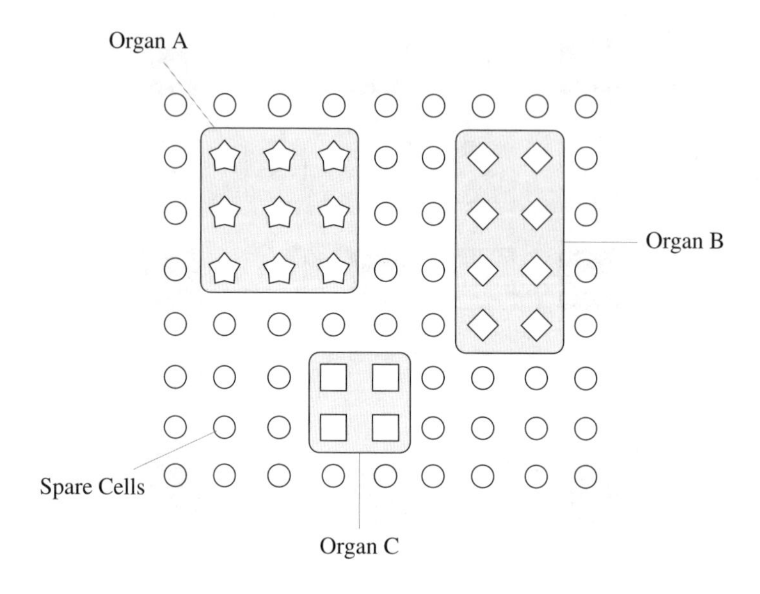

Fig. 8. Architecture due to Greensted et al.

6 Cell Design

Within the research towards the MUXTREE architecture, a fundamental architecture has been established which represents the most typical features of embryonic cells. Based on this design, we present the typical set of features of embryonic cells within this section.

6.1 Embryonic Cell Design

A basic cell architecture based on the MUXTREE design is shown in Figure 9 (based on an image in [27]). This is a cell for a cellular array which employs row deletion in case of an error. In this design, each cell carries the complete genome of the system, i.e., the memory block holds the information for all coordinates. This is a design choice which is not implemented by all cellular arrays. In [23, 5], architectures are shown wherein cells only store the configurations present in their own columns; since only rows can be eliminated in this system, no other knowledge is required by the cells.

The coordinate generator determines which of the memory slots is to be used. For this purpose, it receives the X and Y coordinates located below and left of the cell and increments these coordinates by one. Thus, it is ensured that the cells in the adjacent cells receive the proper coordinates. This mechanism is illustrated by Listing 1 for a matrix with cellular array with dimensions 2 by 3 , taken from [38]. Variables `WX` and `WY` represent coordinate inputs, and `X` and Y are the generated coordinates. Each of the selected *genes* represents one possible cell configuration.

System errors can be selected by diagnosis logic, which works according to the DWC method in this particular cell. The processing element block contains the functional

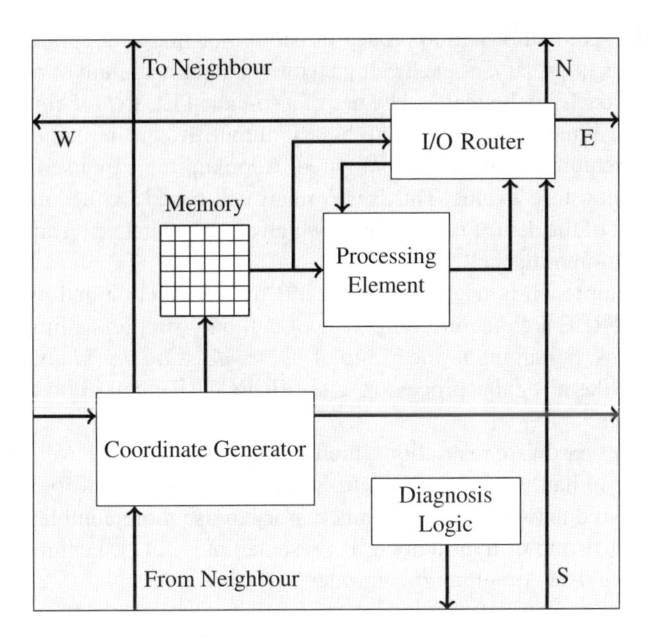

Fig. 9. Design of an Embryonic Cell

block of the cell, explained in detail in section 6.2. The I/O router has the responsibility of making the cell transparent in case of failure.

6.2 Implementation of Functionality

Let us now focus more closely on how the actual calculation functionality is implemented within the device. The most basic function which can be implemented is a two-to-one multiplexer. This type of architecture has been introduced under the name of MUXTREE [18, 29] and has been used in many subsequent designs. By constructing a network of multiplexers, any desired logic function can be implemented. This

Listing 1. Coordinate Generation

```
X = WX+1
Y = SY+1
case of X,Y:
  X,Y = 1,1: do gene A
  X,Y = 1,2: do gene B
  X,Y = 2,1: do gene C
  X,Y = 2,2: do gene D
  X,Y = 3,1: do gene E
  X,Y = 3,2: do gene F
```

is the most fine-grained design approach possible. For practical applications, a more coarse-grained is necessary, as each cell generates a certain amount of overhead.

Such an approach can be realized by use of a so-called LUT (*look-up table*). A look-up table with n binary inputs and one binary output. It can implement 2^n different functions, and requires n bits to be configured. A lookup-table by itself is only combinatoric, it does not store a value. This is why many cell-based architectures also include a register as part of the device functionality, which can be switched on and off according to the configuration of the cell.

A different approach is taken in [33, 19, 39] by introduction and usage of the so-called MICROTREE architecture. This type of cell only supports a limited number of basic instructions, belonging to the group of *binary decision machines*. Thus, it is not programmable like a regular processor. The MICROTREE cell allows to store up to 1024 instructions.

Due to current hardware restrictions, more complex basic cells have not been implemented directly in hardware. However, for implementations meant for the demonstration of bio-inspired networks, it is often necessary to use more complex basic cells. A cell network consisting of hybrid nodes is presented in [11, 12]. Each node consists of a Virtex FPGA and an Atmel microcontroller.

In [1], a virtual network of cells is presented, i.e., it completely runs in software. Each cell is controlled by a configuration format called *eDNA*. This format allows the cell to be programmed with instructions that include control statements such as branches, conditional loops and array operations. A fundamentally different approach is taken by networks of cells which merely perform a function based on the states of their neighbours. The *Game of Life* by Conway is a famous example of such a system. More complex systems of this type, commonly known as *cellular automata*, have been extensively researched.

6.3 Cell Configuration

Early FPGA implementation of bio-inspired cellular arrays used only few bits to decide their configuration. For instance, the design introduced in [27] requires 17 configuration bits, which serve the following purposes:

- Configuration of the cell multiplexer.
- Selection of the proper inputs, outputs and routing.
- Indication whether the cell output is registered.

However, this information is not sufficient in more complex designs, especially if there are no special components external to the cells to handle functions such as cell deletion and error handling. In [43], the memory is split into two sections: While the ordinary configuration data is held in a so-called *DNA segment memory*, the memory for performing additional functions such as error handling is held in a so-called *core register*. In order to illustrate the scope of management functions performed by the cell, we give a description of the flags within its 14 bit wide core register in Table 1.

Table 1. Cell Configuration Memory

Flag	Purpose
Cell Error Status	Flag is set to '1' if the cell has been marked as damaged.
Cell Function Status	Indicate whether the cell is to act as a spare or functional cell.
DNA-Repair Requirement	Set to '1' if the cell needs DNA repair.
DNA-Repair Flag	Indicate that DNA-repair has finished.
Cell-Elimination Requirement	The cell has been irreparably damaged and needs to be eliminated.
Cell-Elimination Flag	Indicate that the cell-elimination requested by previous flag has finished.
Row-Elimination Requirement	The maximum number of cells has been damaged within the current row and it needs to be eliminated.
Row-Elimination Flag	Flag is set to '1' if row-elimination has been completed.
Permission flag of reconfiguration	Indicate whether reconfiguration of the cell is allowed.
Last-Cell Flag	Set to '1' in the last cell of a row.
First-Cell Flag	Set to '1' in the first cell of a row.
Last-Row Flag	Set to '1' in all cells of the last row.
First-Row Flag	Set to '1' in all cells of the first row.
System Error Flag	Indicate that the system cannot be repaired.

7 Technological Requirements for Self-healing Structures

Enabling self-healing mechanisms within a system requires the system to be able to modify itself. This property can be implemented by technology in different ways: On the one hand, software-based implementations may initialize the execution of another program in the course of self-healing actions. On the other hand, hardware-based implementations may make use of the capacity to reconfigure. Thus, not the fundamental structure of a device itself is changed, but for example the wiring of its connection elements and the assignment of its look-up tables which implement logical functions is altered. The most common representative of this architecture type is FPGA technology. Furthermore, less complex devices with these abilities are CPLDs *(complex programmable logic devices)*.

Dynamic partial reconfiguration is a powerful technology for reconfigurable hardware devices with self-healing abilities: Using this paradigm, the configuration of only a part of the device is modified during operation, notably without affecting the remaining areas. The area to be reconfigured is disconnected from the surrounding system, reconfigured and then reintegrated into the system. Thus, in case of a defect or an error, it is possible to repair single cells of the cellular designs.

There exist reconfiguration strategies for both transient and persistent errors of a device. A sudden and unwarranted change in a register value caused by cosmic radiation, referred to as *single event upset*, can be repaired by dynamic partial reconfiguration as registers are reinitialized during the reconfiguration process. In case of a persistent defect within the device which cannot be reverted, the affected area may also be reconfigured.

Because of the generic structure of a reconfigurable device such as an FPGA, it is likely that not all elements contained within the device are used by a configuration. A different hardware mapping may make use of different hardware areas. Therefore, in case of an error, the device may simply retain the original functionality, but switch to a different mapping thereof. If defective area is not used by the new configuration, the device can resume faultless operation. However, to apply this approach, different mappings have to be stored for each functionality.

Using the mechanisms presented in this survey, the ability of a biological organism to replace defective parts can be emulated. However, there are some major differences: Contrary to the concept of dynamic partial reconfiguration applied to a cellular-based approach, a biological organism usually does not try to repair defective cells. They are discarded and replaced by new ones. Furthermore, an organism is able to produce virtually any number of new cells whereas a electronic device is bound to its already existing resources. The resources available on a device are further decreased by faults and subsequent healing actions which render non-repairable areas inactive. In conclusion, the resource bounds place strong limits on the analogy between the mode of operation of biological organisms and of electronic devices.

8 Design Methodologies for Cellular Arrays

Within this article, we have presented several layouts of cellular arrays with self-healing abilites. This section adds remarks on quality and criteria of self-healing functionalities, on different paradigms when designing cellular arrays, and on limitations of the analogy to biology.

8.1 Measuring the Quality of Self-healing Cellular Arrays

When designing a bio-inspired cellular array, several considerations regarding the quality of its self-healing abilites have to be considered. With regard to this question, support is provided by the work presented in [21], which establishes attributes for the quality of self-healing systems, expanding the ISO 9126 quality model. These five additional attributes may be transferred to cellular arrays:

- **Support for Detection of Anomalous System Behavior:** The ability to monitor, recognize and address anomalies within the system. In cellular arrays, monitoring functions cells should be either integrated in each cell itself or in an overlying control instance, a design called the "aggregator-escalator-peer architectural style" in [21]. This has the advantage that a monitoring system can evaluate the state of more than one cell at the same time.

- **Support for Failure Diagnosis:** The ability to locate a fault. In case of cellular arrays, the cell which malfunctions has to be found and addressed.
- **Support for Simulation of Expected or Predicted Behavior:** Based on a framework that allows simulation. Since most of the presented cellular arrays are based on FPGAs, existing design tools already feature the use of mighty system simulators.
- **Differentiating Between Expected and Actual Behavior:** This type of self-awareness uses results from simulation as foundation to detect variations of the system behavior. Cell-based systems can contain a model of the desired behavior in each cell and compare it to its actual behavior to integrate this ability.
- **Support for Testing of Correct Behavior:** With regard to cellular arrays, verification of the entire system can be achieved, as each cell is individually monitored.

8.2 Design and Programming Paradigms

Though all of the concepts presented in this paper rely on a cellular layout of homogeneous cells, the inner structure of cells may widely differ as seen in Section 6. Moreover, not only the inner cell design, but also the overlying design and programming paradigms vary.

For instance, in the eDNA approach published in [1], the cells, referred to as *eCells*, are connected through a network-on-chip. The *eDNA* within each cell defines the functionality. An eCell resembles to a CLB (configurable logic block) known from FPGAs, but also has a small processing unit. Each eCell is bound to a definite identifier. The destination of data of an output port of an eCell is determined based on these identifiers.

The advantage of this approach is the possibility to describe algorithms for it in a high-level language. Parallel sections have to be marked explicitly, and thus, the massive-parallel layout of the cellular array can be utilized. Each of the operators used in this language is assigned a number. In an intermediate step, the code is translated to be executable on the hardware. Each sequence of code with a number is assigned to a different cell. A dedicated start signal on each eCell based on the number scheme organizes the controlled sequencing. This process is considered to be a self-organization process. In case of a faulty cell, an eCell detected as defective is marked as dead.

The approach assumes BIST or other failure detection mechanisms. The identifier of a dead cell is deleted and assigned to the cell with the identifier next in size. Therefore, all subsequent eCells are assigned new identifiers and the process of self-organization has to be restarted. However, this type of self-healing affects large parts of the running systems since there is need for a new self-organization process. This approach has the advantage of using a high-level language to describe a parallel cellular-based design with self-healing abilities.

The MUXTREE architecture, described in section 5.1 also features programmability using a high level language. A cell in the MUXTREE approach consists of a 2-1 multiplexer which chooses between two 8-1 multiplexers. These deliver either constants, values from neighbors or a saved result from the cell's register at their inputs. Neighboring cells are connected via a switch matrix similar to those found in FPGAs. The configuration of the multiplexers are stored within a *gene*. To program this architecture, the language NANOPASCAL was developed. It consists of simple commands which are either conditional statements or assignments, but does not allow branches. A compiler is available which computes the target address of the genes, i. e. the configuration bits of each cell. Related to this MUXTREE implementation is the realization of the MUXTREE architecture as a set of universal Turing machines as shown in [31]. In addition, there is the programming language PICOPASCAL, which has a similar structure as NANOPASCAL.

8.3 Cell Granularity

Many of the approaches presented in this survey share the same fundamental structure: A two-dimensional array of homogeneous function units, which are known as cells, constitute the bio-inspired systems. However, the functionality of the cells may vary significantly. In [19], a functional unit consists of three multiplexers and a register. The authors describe this as the "molecular" level. Functional units can be combined into a *microtree*, a small processing unit resembling a CPU, on the "cellular" level.

As in [31], cells are considers as complete universal Turing machines. This approach is based on the work of the mathematician John von Neumann in work first published in 1966 [22]. Depending on the granularity of the inner cell structure, programming paradigm differ. Whereas for very fine-grained cells, special programming languages and their interpreters or compilers often need to be defined, cellular arrays with cells as complex as CPUs possibly resemble common multiprocessor architectures. Therefore, existing paradigms for multiprocessor programming could be adapted to these systems.

8.4 Evaluation of the Analogy

An essential decision while designing bio-inspired systems is the question which concepts and mechanisms known from biology are worthy of adoption. Depending on the intended use, either a design emulating DNA-based cell and immune system is considered as a bio-inspired system or rather as a composition of homogeneous functional units with self-replicating abilities. The amount and complexity of biological concepts which have been adopted, however, allows no conclusion with regard to the quality of a bio-inspired system: The concept of a central nervous system, for example, is indeed simplified imaginable as a central processing unit within the design which covers monitoring and control functions. However, for living organisms, this central system forms a critical element. In case of irreversible injury, the organism will eventually die or rather not be able to continue autonomous operation. Within the class of electronic devices, a system with such a central unit would pose a single point of failure. Therefore, it has to be evaluated if bio-inspired systems should prefer a more decentralized approach to execute the tasks mentioned above, in spite of the desired analogy to biology.

Furthermore, while many of the authors of papers based on bio-inspired systems use many terms defined in the field of biology, this analogy is misleading in some aspects. For example, the term *cicatrization* is used in [19] to advertise that a defective cell has been replaced and the function running within has performed a reset. Apart from the modified cell and system state, the new cell is identical with regard to its functionality in comparison to the original cell. In terms of biology, however, cicatrization means the creation of inferior tissue as replacement for tissue irreversibly damaged after an injury.

Additionally, different options exist with regard to the functionalities held ready within each cell. In some approaches, a single cell only has the functionality needed to execute its own task; this corresponds to fully differentiated cells in biology. If one of these cells fails, surrounding cells are not able to resume their tasks.

In other approaches, as for example in [35], cells also contain functionality of their neighboring cells. In biology, this corresponds to stem cells which are able to further differentiate to several different types of cells. This causes a higher resource consumption within each cell, nevertheless, in case of a fault of a cell, a neighboring cell may replace the defective cell. Further approaches could include the complete functionality of a system within each cell. This corresponds to zygotic cells which are able to differentiate to every type of cells of an organism. Thus, depending on the overall functionality of the system and the granularity of the cells, resource consumption for each cell can be huge. However, this may eventually allow the system to restore its original functionality even in case of multiple cell defects. Thus, not every concept known from biology is always suited to be transferred to electronic systems and in some cases, the analogy seems to be overused. The cellular array approaches presented in this paper have the ability of self-replication and self-healing through reconfiguration and sometimes even immune system-like designs like in [15]. They show in numerous publications that biological concepts can be adapted to electrical systems in a reasonable manner.

9 Conclusion and Future Outlook

Biologically-inspired systems are modeled on living organisms and strive to adapt concepts known from nature. Especially in the field of self-healing systems, biology may contribute several aspects: A structure based on cells which can easily be replaced in case of faults, simulated growth or mutation of structures by using reconfigurable devices are ideas which are borrowed from nature.

However, this analogy is often overused or used imprecisely. In contrast to a biological organism, resources of electronic devices are always bounded. The number of different functionalities which can be run on a system at the same time is limited thereby. While an organism usually replaces defective cells without any limitation, faults in a electronic device are often only repaired by using redundant or spare resources. The use of dedicated spare cells which are only activated in case of faults further intensifies the scarcity of resources.

Despite the ability of many of the presented biologically-inspired systems to tolerate or even to recover from faults or attacks, only few of them try to adopt the concept of an immune system, for example [2, 3]. Only if immune systems have been adapted, it will be possible for a bio-inspired system to be protected against a new appearance of the same fault or attack. In order to “educate” artificial immune systems, concepts used in machine learning could be considered. Using machine learning, the epigenesis axis could further be explored.

Even if all of these capabilities would be combined, it remains unclear how to apply existing paradigms for programming such architectures. Given the numerous existence of homogeneous cells in most of the presented approaches, paradigms like CUDA or OpenCL, known from the field of massively parallel devices, could be employed. Some approaches, such as [1] or [18], already deliver new mechanisms to use constructs known from common programming languages.

In summary, despite still being a niche in the area of self-healing systems, bio-inspired systems not only cover aspects known from nature, but also offer the possibility to combine different fields of research such as design methodologies for massively parallel devices or machine learning. Possible fields of application are fully autonomous systems which have to be highly reliable, such as nodes in isolated sensor networks and devices exposed to cosmic radiation.

References

1. Boesen, M., Madsen, J.: eDNA: A Bio-Inspired Reconfigurable Hardware Cell Architecture Supporting Self-organisation and Self-healing. In: 2009 NASA/ESA Conference on Adaptive Hardware and Systems, pp. 147–154 (2009)
2. Bradley, D., Tyrrell, A.: Hardware fault tolerance: An immunological solution. In: IEEE International Conference on Systems, Man, and Cybernetics, vol. 1, pp. 107–112. Citeseer (2000)
3. Bradley, D., Tyrrell, A.: The architecture for a hardware immune system, p. 0193 (2001)
4. Canham, R., Tyrrell, A.: A learning, multi-layered, hardware artificial immune system implemented upon an embryonic array. In: Tyrrell, A.M., Haddow, P.C., Torresen, J. (eds.) ICES 2003. LNCS, vol. 2606, pp. 174–185. Springer, Heidelberg (2003)
5. Canham, R., Tyrrell, A.: An embryonic array with improved efficiency and fault tolerance. In: Proceedings of the 2003 NASA/DoD Conference on Evolvable Hardware EH 2003, p. 275. IEEE Computer Society, Washington (2003)
6. Canham, R., Tyrrell, A.: A hardware artificial immune system and embryonic array for fault tolerant systems. Genetic Programming and Evolvable Machines 4(4), 359–382 (2003)
7. Dasgupta, D., Ji, Z., Gonzalez, F., et al.: Artificial immune system (AIS) research in the last five years. In: Proceedings of The 2003 Congress on Evolutionary Computation (CEC 2003), pp. 123–130 (2003)
8. Forrest, S., Perelson, A., Allen, L., Cherukuri, R.: Self-nonself discrimination in a computer. In: Proceedings of IEEE Computer Society Symposium on Research in Security and Privacy, pp. 202–212 (1994)
9. Gericota, M., Alves, G., Ferreira, J.: A self-healing real-time system based on run-time self-reconfiguration. In: 10th IEEE Conference on Emerging Technologies and Factory Automation, ETFA 2005, vol. 1 (2005)
10. Ghosh, D., Sharman, R., Raghav Rao, H., Upadhyaya, S.: Self-healing systems–survey and synthesis. Decision Support Systems 42(4), 2164–2185 (2007)
11. Greensted, A., Tyrrell, A.: An endocrinologic-inspired hardware implementation of a multicellular system. In: NASA/DoD Conference on Evolvable Hardware, Seattle, USA (2004)
12. Greensted, A., Tyrrell, A.: Implementation results for a fault-tolerant multicellular architecture inspired by endocrine communication. In: Proceedings of NASA/DoD Conference on Evolvable Hardware, pp. 253–261 (2005)
13. Hammerstrom, D.: A survey of bio-inspired and other alternative architectures. In: Nanotechnology. Information Technology II, vol. 4 (2008)
14. Lala, P., Kumar, B.: An architecture for self-healing digital systems. In: Proceedings of the Eighth IEEE International On-Line Testing Workshop, pp. 3–7 (2002)

15. Lala, P., Kumar, B.: Human immune system inspired architecture for self-healing digital systems. In: Proceedings of International Symposium on Quality Electronic Design, pp. 292–297 (2002)
16. Macias, N., Athanas, P.: Application of Self-Configurability for Autonomous, Highly-Localized Self-Regulation. In: Second NASA/ESA Conference on Adaptive Hardware and Systems, AHS 2007, pp. 397–404 (2007)
17. Mange, D., Goeke, M., Madon, D., Stauffer, A., Tempesti, G., Durand, S.: Embryonics: A new family of coarse-grained field-programmable gate array with self-repair and self-reproducing properties. In: Towards evolvable hardware, pp. 197–220 (1996)
18. Mange, D., Sanchez, E., Stauffer, A., Tempesti, G., Marchal, P., Piguet, C.: Embryonics: a new methodology for designing field-programmable gatearrays with self-repair and self-replicating properties. IEEE Transactions on Very Large Scale Integration (VLSI) Systems 6(3), 387–399 (1998)
19. Mange, D., Sipper, M., Stauffer, A., Tempesti, G.: Towards robust integrated circuits: The embryonics approach. Proceedings of the IEEE 88(4), 516–541 (2000)
20. Neti, S., Muller, H.: Quality criteria and an analysis framework for self-healing systems. In: ICSE Workshops International Workshop on Software Engineering for Adaptive and Self-Managing Systems SEAMS 2007, p. 6 (2007)
21. Neti, S., Muller, H.A.: Quality criteria and an analysis framework for self-healing systems. In: International Workshop on Software Engineering for Adaptive and Self-Managing Systems, p. 6 (2007)
22. Neumann, J.V.: Theory of Self-Reproducing Automata. University of Illinois Press, Champaign (1966)
23. Ortega, C., Tyrell, A.: MUXTREE revisited: Embryonics as a reconfiguration strategy in fault-tolerant processor arrays. In: Sipper, M., Mange, D., Pérez-Uribe, A. (eds.) ICES 1998. LNCS, vol. 1478, p. 217. Springer, Heidelberg (1998)
24. Ortega, C., Tyrrell, A.: Reliability analysis in self-repairing embryonic systems. Memory 12, 11 (1999)
25. Ortega, C., Tyrrell, A.: Self-repairing multicellular hardware: A reliability analysis. In: Floreano, D., Mondada, F. (eds.) ECAL 1999. LNCS, vol. 1674, pp. 442–446. Springer, Heidelberg (1999)
26. Ortega, C., Tyrrell, A.: A hardware implementation of an embryonic architecture using Virtex® FPGAs. In: Miller, J.F., Thompson, A., Thompson, P., Fogarty, T.C. (eds.) ICES 2000. LNCS, vol. 1801, pp. 155–164. Springer, Heidelberg (2000)
27. Ortega, C., Tyrrell, A.: A hardware implementation of an embryonic architecture using virtex fpgas. In: Miller, J.F., Thompson, A., Thompson, P., Fogarty, T.C. (eds.) ICES 2000. LNCS, vol. 1801, pp. 155–164. Springer, Heidelberg (2000)
28. Ortega-Sanchez, C., Mange, D., Smith, S., Tyrrell, A.: Embryonics: A bio-inspired cellular architecture with fault-tolerant properties. Genetic Programming and Evolvable Machines 1(3), 187–215 (2000)
29. Ortega-Sanchez, C., Tyrrell, A.: Design of a basic cell to construct embryonic arrays. IEE Proceedings-Computers and Digital Techniques 145(3), 242–248 (1998)
30. Prodan, L., Tempesti, G., Mange, D., Stauffer, A.: Embryonics: electronic stem cells. Artificial life eight, 101 (2003)
31. Restrepo, H.F., Mange, D.: An embryonics implementation of a self-replicating universal turing machine, pp. 74–87 (2001)
32. Sanchez, E., Mange, D., Sipper, M., Tomassini, M., Pérez-Uribe, A., Stauffer, A.: Phylogeny, ontogeny, and epigenesis: Three sources of biological inspiration for softening hardware. In: Higuchi, T., Iwata, M., Weixin, L. (eds.) ICES 1996. LNCS, vol. 1259, pp. 33–54. Springer, Heidelberg (1997)

33. Stauffer, A., Mange, D., Goeke, M., Madon, D., Tempesti, G., Durand, S., Marchal, P., Piguet, C.: MICROTREE: Towards a Binary Decision Machine-Based FPGA with Biological-like Properties, pp. 103–112 (1996)
34. Szasz, C., Chindris, V.: Artificial life and communication strategy in bio-inspired hardware systems with FPGA-based cell networks. In: 11th International Conference on Intelligent Engineering Systems, INES 2007, pp. 77–82 (2007)
35. Szasz, C., Chindris, V.: Development strategy and implementation of a generalized model for FPGA-based artificial cell in bio-inspired hardware systems. In: 5th IEEE International Conference on Industrial Informatics, vol. 2 (2007)
36. Szasz, C., Chindris, V.: Bio-inspired hardware systems development and implementation with FPGA-based artificial cell network. In: IEEE International Conference on Automation, Quality and Testing, Robotics, AQTR 2008, vol. 1 (2008)
37. Szasz, C., Czumbil, L.: Artificial molecule development model for genes implementation in bio-inspired hardware systems. In: 11th International Conference on Optimization of Electrical and Electronic Equipment, OPTIM 2008, pp. 15–20 (2008)
38. Tempesti, G., Mange, D., Mudry, P., Rossier, J., Stauffer, A.: Self-replicating hardware for reliability: The embryonics project. ACM Journal on Emerging Technologies in Computing Systems (JETC) 3(2), 9 (2007)
39. Tempesti, G., Mange, D., Stauffer, A.: A self-repairing FPGA inspired by biology. In: Proc. 3rd IEEE Int. On-Line Testing Workshop, pp. 191–195 (1997)
40. Tempesti, G., Mange, D., Stauffer, A.: Self-replicating and self-repairing multicellular automata. Artificial Life 4(3), 259–282 (1998)
41. Tyrrell, A., Sanchez, E., Floreano, D., Tempesti, G., Mange, D., Moreno, J., Rosenberg, J., Villa, A.: Poetic tissue: An integrated architecture for bio-inspired hardware. In: Tyrrell, A.M., Haddow, P.C., Torresen, J. (eds.) ICES 2003. LNCS, vol. 2606, pp. 269–294. Springer, Heidelberg (2003)
42. Yao, X., Higuchi, T.: Promises and challenges of evolvable hardware. In: Evolvable Systems: From Biology to Hardware, pp. 55–78 (1997)
43. Zhang, X., Dragffy, G., Pipe, A., Gunton, N., Zhu, Q.: A reconfigurable self-healing embryonic cell architecture. Differentiation 1, 4 (2003)

Combined Man-in-the-Loop and Software-in-the-Loop Simulation

Electronic Stability Program for Trucks on the Daimler Driving Simulator

Uwe Baake and Klaus Wüst

Daimler AG
Company Postal Code B209
70546 Stuttgart, Germany
{uwe.baake,klaus.wuest}@daimler.com

Abstract. The main targets in commercial vehicle development in the near future will be improving the energy effiency of the vehicles and improving vehicle safety. One of the measures to increase safety is the decision of the European Committee to make electronic stability systems compulsory for nearly all trucks and buses. To guarantee that the system performs well for a wide variety of trucks and buses, new simulation methods are being introduced into the development process. The system functionalities, which are developed by system suppliers, are implemented by Daimler Trucks as software-in-the-loop codes into vehicle dynamics simulation models. By using the multi-body simulation software's real-time capabilities, it has become possible to investigate the interaction between the vehicle and the electronic stability system on the Daimler driving simulator.

Keywords: Driving simulator, Electronic Stability Program, Software-in-the-Loop, Man-in-the-Loop, Multi-body systems simulation, Commercial vehicle development.

1 Introduction

The transportation of people and goods plays a key role in the globally networked economy and is indispensable for the growth and prosperity of every society. Goods transportation is continually expanding, and road transport in particular is registering high growth rates. Experts are predicting an annual global increase in transportation volumes in the order of 2.5 percent on average by the year 2030; this represents double the figure for the year 2000. Consequently, development of commercial vehicles is facing two major challenges for the near future [1], grouped together in the *Shaping Future Transportation* initiative as shown in Fig. 1.

Shaping future transportation means preserving resources and reducing emissions of all kinds, while at the same time ensuring maximum road safety. The

A. Biedermann and H. Gregor Molter (Eds.): Secure Embedded Systems, LNEE 78, pp. 171–185.
springerlink.com

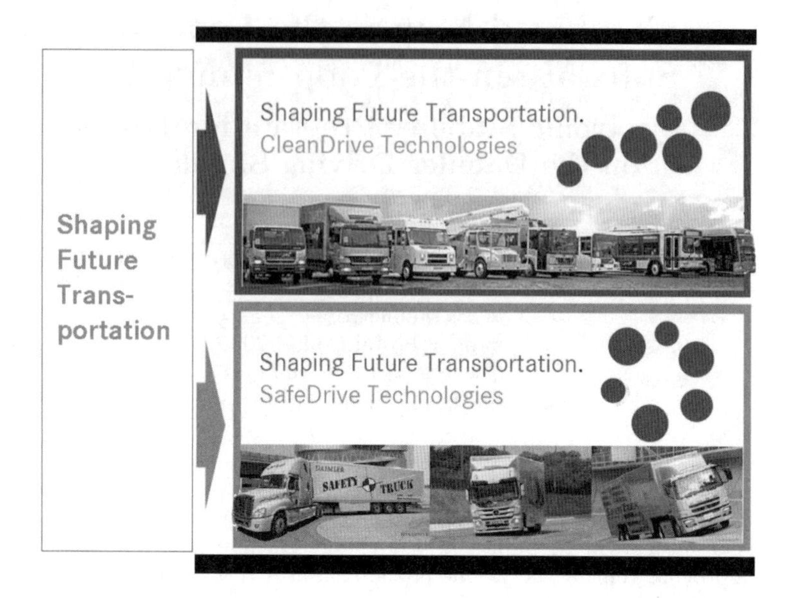

Fig. 1. Daimler technology programs

Clean Drive Technologies program has been initiated in order to offer new products with overall operating efficiency but also with low emissions and low fuel consumption. Within this program, the focus of all efforts is on fuel consumption and thus indirectly on reducing CO_2 emissions, as demanded by factors such as increasing fuel prices and increasing emission standards. This will require optimization of the conventional drive trains as well as the introduction of alternative drive train systems such as hybrid drives. The work done so far has led to fuel consumption and the emissions limited by regulations (NO_X, HC, CO, particulate matter) being drastically reduced for all commercial vehicles. Recently, with many commercial vehicles having already fulfilled EURO V and since the announcement by the European Commission in 2006 that emission limits would be made much stricter once again, the industry's focus has been on fulfilling the emission limits of the upcoming EURO VI legislation.

The goal of accident reduction thus remains ambitious and challenging and is the main focus of the second program, *Safe Drive Technologies*, as outlined in Fig. 1. Road safety is influenced by numerous factors. Road user behavior, weather-related influences, vehicle technology, traffic infrastructure, and efficient emergency services all play a role. Due in particular to the improvements of vehicle safety technology, the number of fatalities on German roads each year has fallen continuously since 1970 and reached a historic low of 4,949 in 2007.

It is equally true for Germany and the whole of Europe that the number of vehicles on the roads has approximately tripled and mileages have approximately doubled during that period. This means that the risk of an accident – the accident rate measured by the number of accidents in relation to total vehicle kilometers driven – has fallen by more than 90% since 1970. The positive

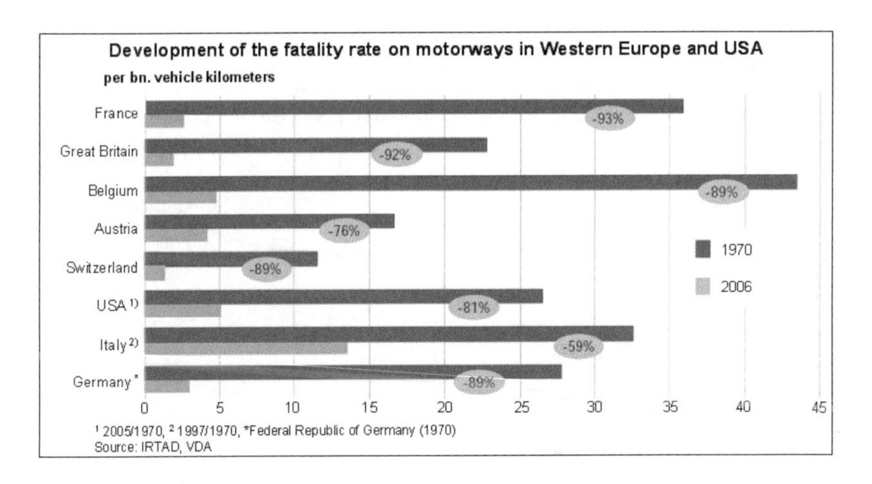

Fig. 2. Development of the fatality rate on motorways in Western Europe and USA

results of road safety work in Germany are also reflected by the development of the accident rate on the autobahns. Compared internationally, Germany has some of the best statistics. The probability of a fatal accident on German autobahns is now only one third as high as on other roads. The enormous road safety progress made by commercial vehicles is impressively demonstrated by accident statistics. Although goods transport on the roads has risen by 71% since 1992, the number of fatal accidents involving trucks has decreased by 36% during the same period (see Fig. 3). When these figures are seen statistically in relation to the increased transport volumes, the accident rate has actually fallen by 63%. Substantial road safety progress has been achieved through the deployment of innovative active safety systems in commercial vehicles. Besides the development of driver assistance systems, one of the measures to achieve the goal of increasing commercial vehicle safety is the introduction of the *Electronic Stability Program* (ESP) for trucks and buses. ESP is very common in passenger cars nowadays and has proven to be an essential factor in reducing traffic accidents. For commercial vehicles, electronic stability systems have been available in series production vehicles since 2001. By modifying the European Committee (ECE) regulation for the braking systems of heavy commercial vehicles, the European Union has started an initiative for making ESP compulsory for nearly all trucks and buses (except construction trucks and city buses), with a schedule starting in 2011.

A survey conducted by the Mercedes-Benz passenger car division in 2002 has shown that the evaluation of accidents with Mercedes-Benz passenger cars clearly proves the ability of ESP to reduce accidents. After the introduction of ESP for all Mercedes-Benz passenger cars in 1999, the number of accidents was reduced significantly already in the years 2000 and 2001 as outlined in Fig. 4 [2]. It is expected that a similar effect on accident reduction may be achieved by the mandatory introduction of ESP for trucks and buses.

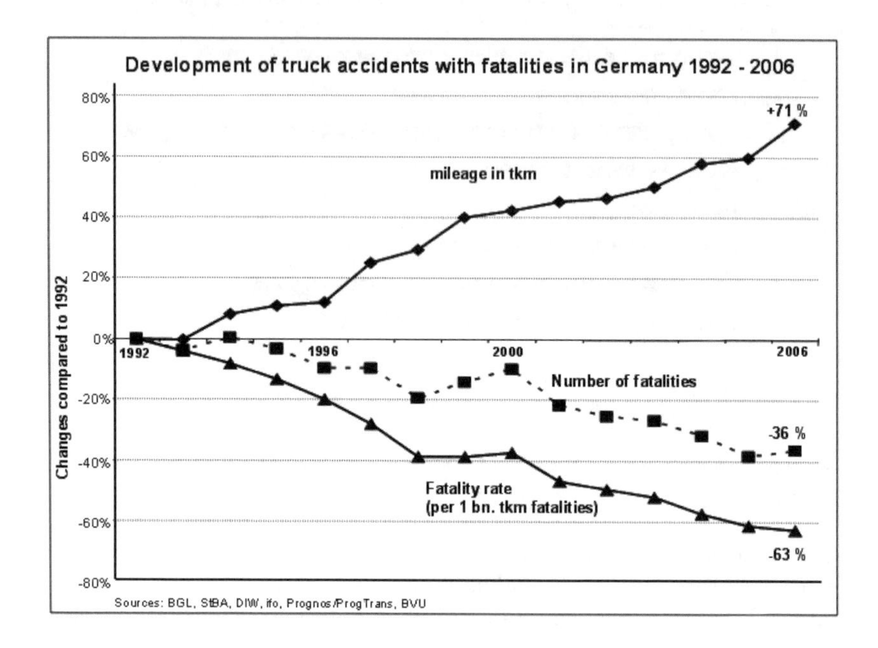

Fig. 3. Development of truck accidents with fatalities in Germany 1992 – 2006

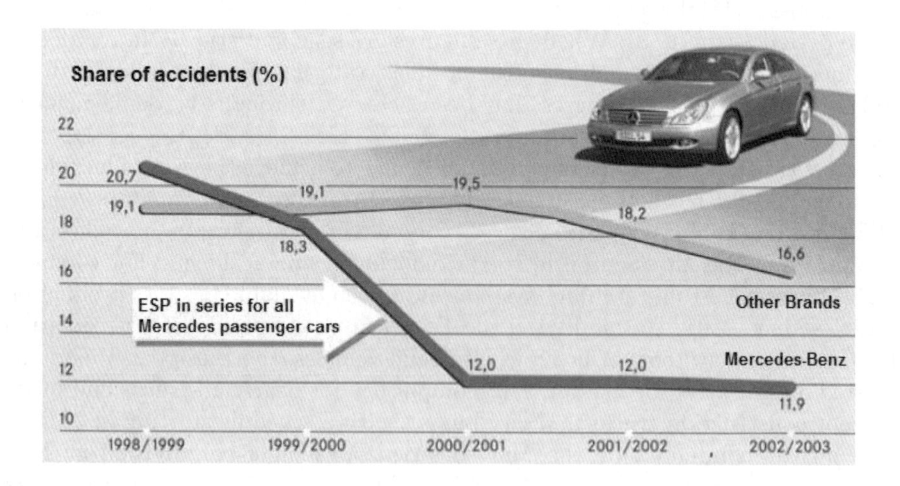

Fig. 4. Decrease of accidents due to ESP

2 ESP for Commercial Vehicles

2.1 System Description

ESP for trucks and buses has two basic functionalities, *Electronic Stability Control* (ESC) and *Rollover Control* (RSC). ESC works in a similar way to that in passenger car systems. Sensors are measuring steering wheel angle, yaw velocity and lateral acceleration and feed the values into a simple vehicle model. As soon as this model detects a big difference between the desired and the measured reaction of the vehicle upon the steering wheel input, either in an oversteering or an understeering direction, braking forces are applied to individual wheels generating a moment around the vehicle's vertical axis, which stabilizes the vehicle movement as shown in Fig. 5.

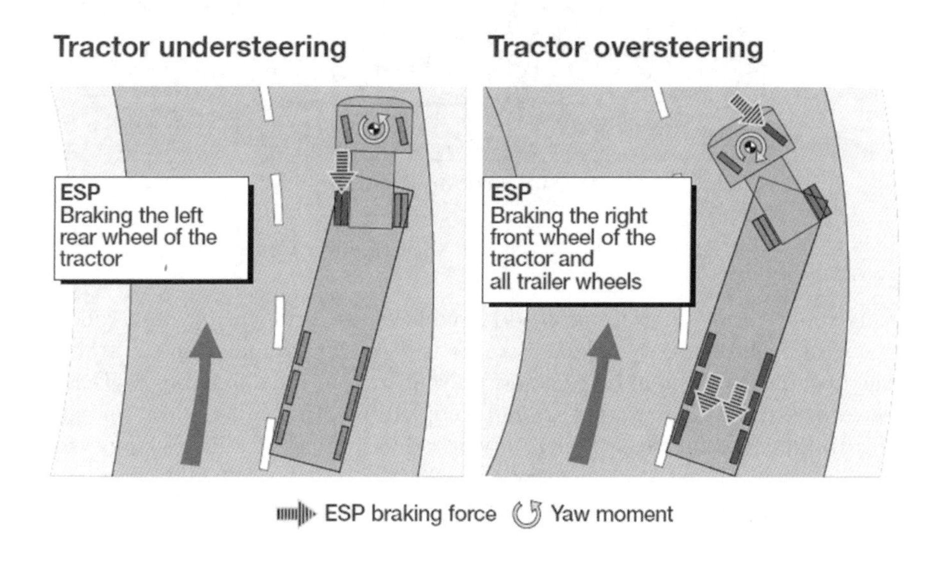

Fig. 5. ESC as an ESP Functionality

For commercial vehicles, especially when laden, limits of adhesion are only reached on roads with a low coefficient of friction. For a fully laden truck with an elevated load, rollover may already occur for steady-state lateral accelerations below 4 m/s^2, which is far below the limits of adhesion of the tyres on dry road surfaces. For trucks and buses travelling on dry road surfaces, it is therefore more important to prevent them from reaching the rollover limit. The rollover control limits the lateral acceleration of the vehicle to a pre-defined value. As the rollover limit of a commercial vehicle strongly depends on the loading condition of the vehicle, a mass detection is integrated into the rollover control, which adapts the intervention threshold value of lateral acceleration to the vehicle mass and additionally to the vehicle speed. When the vehicle reaches this load- and speed-dependent maximum value of lateral acceleration, all wheels of the

(a) Without Rollover Control

(b) With Rollover Control

Fig. 6. Truck without and with Rollover Control

towing vehicle and trailer are braked to reduce speed and thus prevent rollover. Braking of both towing vehicle and trailer is necessary to maintain the stability of the tractor/trailer combination and prevent it from the dangerous *jackknifing* effect, where the trailer swerves around completely until it hits the tractor cabin.

For Daimler trucks, the system is integrated into the electronic braking system and has been available on the market since 2001 as *Telligent Stability System*, integrated into a safety package together with other systems such as *Lane Keeping Assistant* and *Active Cruise Control* (ACC).

Since 2005, these safety packages for trucks and buses also include *Active Brake Assist*. With the ACC sensors, this system detects if the truck or bus is quickly approaching a car which is driving ahead and is decelerating by exceeding the maximum deceleration of ACC. In a first step, the system warns the driver with an acoustic signal. If despite this warning there is no reaction by the driver, the system activates the braking system and brakes at first with a moderate deceleration and then with fully applied brakes until standstill. The main benefit of this system is the avoidance of accidents, when a truck or bus approaches the end of a traffic jam.

However, until now these systems have only been available for tractor/semi-trailer combinations and coaches as outlined in Fig. 7. Mainly because of the additional costs for the operator, the percentage of vehicles equipped with the system is still low, although there is not only the advantage of reduced accidents, but also many insurance companies are offering reductions for vehicles equipped with safety systems.

(a) Safety Truck

(b) Safety Coach

Fig. 7. Mercedes-Benz Actros *Safety Truck* and Mercedes-Benz Travego *Safety Coach*

2.2 System Development

Because of aforementioned reasons, the legislation committee of the European Union has decided to make ESP systems for trucks and buses compulsory. Other systems like the *Lane Keeping Assistance* and the *Active Brake Assist* will probably follow. The newly-defined ECE-R13 legislation [3] including mandatory ESP systems will come into effect in 2011, starting with ESP systems for coaches and for tractor/semitrailer combinations, where it is already available. All other trucks and buses except vehicles with more than 3 axles and vehicles for construction purposes will have to follow by 2014.

This makes it necessary for the truck and bus manufacturers and the braking system suppliers to develop the system for a wide variety of commercial vehicles. As the amount of vehicles which can be used for proving the system functionalities in field testing is limited, it is necessary to use vehicle dynamics simulation to support the development of the system for the whole variety of commercial vehicle parameters, such as different axle configurations, different wheelbases, tyre variations and various loading conditions as shown in Fig. 8.

(a) Field testing

(b) Simulation

Fig. 8. ESP field testing and ESP simulation

Furthermore, the fact that the system must be designed to cover the whole range of commercial vehicles requires a flexible ESP system, which can be parameterized and which is able to detect not only the loading conditions but also the overall, present vehicle behavior. For this purpose, parameters like the wheelbase, the axle configuration and the steering ratio can be fed into the system during the production process, and additionally the system is able to detect the self-steering gradient of the vehicle and gets additional information during driving, for example the status of lift axles.

Figure 9 shows the simulation approach deployed for ESP development. Field testing results of few vehicle types are used to validate the basic simulation model, which is at first parameterized according to the test vehicle. The results of the simulation concerning the vehicle's dynamic performance and the functions of the ESP system both have to show a good correlation to the measurements. For this model validation, internationally standardized driving maneuvers are used both in simulation and field testing, for example steady-state cornering or lane change maneuvers [4,5]. On the basis of a successful validation, the simulation

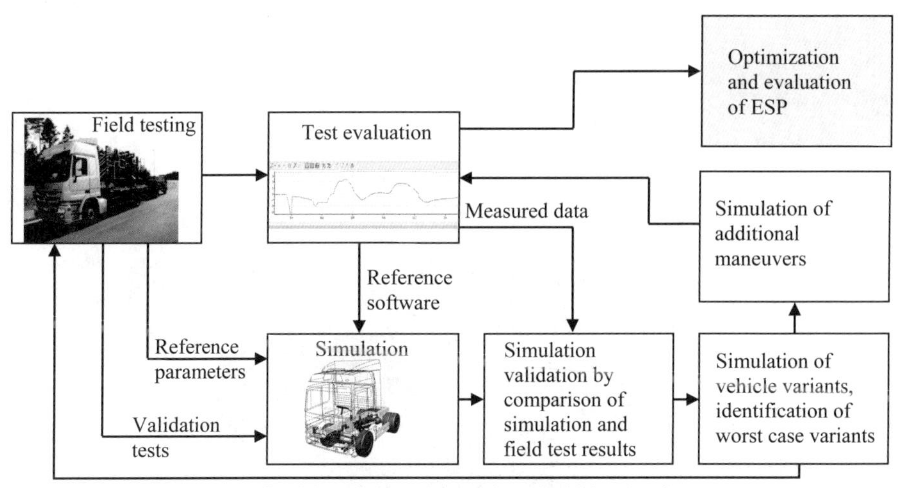

Fig. 9. Simulation-assisted ESP development

is then used to cover the whole range of vehicle variations (by varying vehicle parameters inside the model) and to cover the whole range of possible driving situations (by varying the maneuvers inside the simulation process).

2.3 Software-in-the-Loop Integration of ESP into Vehicle Models

To investigate the vehicle behavior with ESP in simulation, a code of the system has to be integrated into the vehicle dynamics simulation tool. The *Computer-Aided Engineering* (CAE) analysis division of Daimler Trucks uses *Simpack* as a standard multibody simulation tool for different purposes. This software package offers multiple possibilities for the integration of system codes into the simulation [6].

With the aid of different sets of parameters inside the tyre model, either a dry road surface or a low friction road can be simulated. As the tyre is the only contact between the vehicle and the road, the modeling of the tyre is very important for a realistic vehicle dynamics simulation. For the purpose of ESP simulation with Simpack, MFTYRE is used as a standard tool for tyre modeling [7]. The parameters of the tyre model are gained with measurements conducted either on tyre test stands or on tyre road testing devices as outlined in Fig. 11.

For the integration of truck ESP into vehicle simulation models, Daimler and the system suppliers have chosen MATLAB/Simulink as an exchange and integration platform. The CAE analysis division uses the Simpack code export feature to convert the vehicle simulation model into a transferrable code, which is combined with an S-function of the ESP code generated by the system

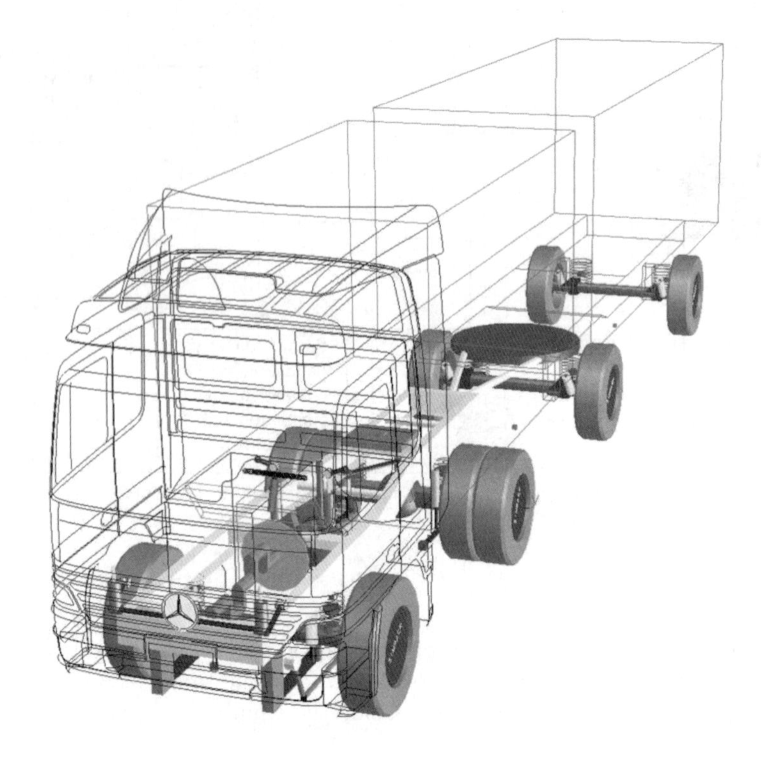

Fig. 10. Truck simulation model

Fig. 11. Tyre test stand

supplier. The definition and evaluation of the various driving maneuvers which have to be investigated to prove the functionality of the ESP system (for example, lane change maneuvers on different road surfaces or closing curve maneuvers) is generated within MATLAB/Simulink as shown in Fig. 12.

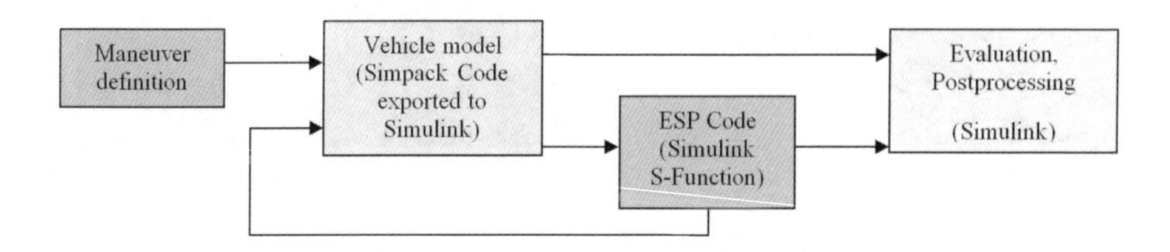

Fig. 12. Simulink simulation with exported Simpack model and ESP

3 Man-in-the-Loop Simulation: The Driving Simulator

3.1 Transfer of Vehicle Models to the Driving Simulator

The Daimler driving simulator in Berlin was introduced in 1995 and is presently moved to Sindelfingen. As shown in Fig. 13, a complete car or a truck cabin is installed on a hexapod. The hexapod comprises 6 hydraulic cylinders which allow rotational movements of the device around all three coordinate axles. Additionally, the hexapod is fixed on a rail which can be moved in lateral direction [8]. The movements of the simulator are generated by a vehicle simulation model. This simulator serves for investigations of the interactions between the driver and the vehicle. It is partly used for subjective evaluations of driver assistance systems – for example, the *Brake Assist* that guarantees full braking application during emergency situations was developed on the basis of simulator investigations. Other systems which were designed with the aid of the simulator are *Lane Keeping Assistance* and *Active Brake Assist*. The simulator is also used for the subjective evaluation of parameter changes within the chassis layout during the development phase of new passenger cars and commercial vehicles.

The Daimler commercial vehicles CAE analysis division has used the driving simulator since 2006, when the real-time capabilities of the multi-body simulation software Simpack allowed a whole variety of simulation models, which are used for the chassis layout investigations, to be transferred to the driving simulator. Since then, several driving simulator tests have been conducted with Simpack vehicle models, which were used to define the target values for the vehicle dynamics of new truck or van generations and for the subjective evaluation of many chassis variants before field testing.

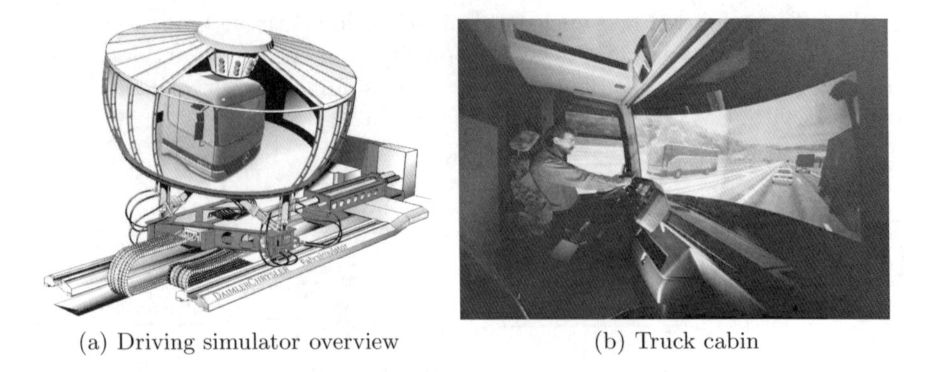

(a) Driving simulator overview (b) Truck cabin

Fig. 13. Daimler driving simulator

The vehicle models used for simulation of the driving behavior have to be detailed enough to guarantee a good representation of the vehicle dynamics behavior. In contrary, they also have to be simple enough to guarantee the real-time capabilities for the driving simulator tests. This goal is achieved by reducing flexible bodies (for example, the frame rails of the truck) to simple rotational spring-/damper-units, by reducing the number of tyre/road contacts (twin tyres are reduced to one tyre) and by using a specially developed equation solver with constant step size (usually 1 ms). The validation of the exported real-time multibody simulation models (see also Chapt. 2.2) is checked within a user interface that allows quick changes of vehicle parameters and driving maneuvers as presented in Fig. 14, before transferring the models to the simulator. Even experienced test drivers agree that these models give a very realistic feel of the dynamic behavior of a truck in the simulator.

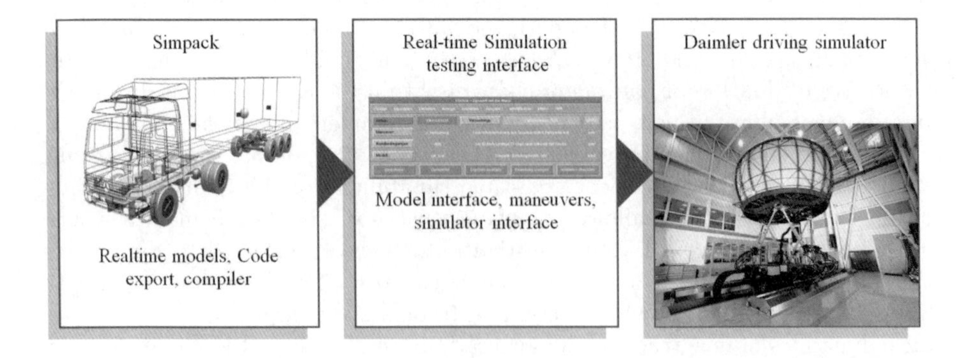

Fig. 14. Transfer of vehicle models to the driving simulator

3.2 ESP Investigations on the Driving Simulator

An initial test with ESP for trucks on the driving simulator was conducted in November 2009. The test served as a basis for investigating if the simulator could give additional value to the simulation-supported development of ESP for trucks. The basic questions were, if the simulator would be able to realistically reproduce the interventions of an ESP system for trucks, and if the simulator could be used for optimizing the system, e.g. for defining the intervention threshold values for different vehicles and various vehicle parameters.

As the Windows-based MATLAB/Simulink software-in-the-loop environment was not suited for the Unix operating system of the driving simulator, it was decided to use a different approach for this first test. A Fortran code, which simulates the basic ESP functionalities of stability control and rollover protection, was integrated into a real-time vehicle model as a user routine. Fortran was chosen, as Simpack offers a standard user routine interface for Fortran routines. The model was then exported into the simulator environment together with the user routine, which also could be parameterized after the code export. With

(a) Lane Change

(b) Slalom

Fig. 15. Lane Change and Slalom maneuver on the driving simulator

this approach, both vehicle parameters and ESP parameters could be varied easily during the simulator test. For example, the threshold values for the intervention of both stability control and rollover control were variable during the test. Beside the parameters of the ESP system, variations of vehicle parameters such as loading conditions, steering parameters and tyre characteristics were also investigated.

Different driving situations were chosen to evaluate the ESP interventions on the driving simulator, all based on straight-ahead driving on a highway track as shown in Fig. 15. Pylons were used for various lane-change and slalom maneuvers with different vehicle velocities. Similar to the validation process of the vehicle models, these maneuvers are based on ISO standards which are applied both in field testing and in simulation [5].

3.3 Results

The main result was to prove that the interventions of the ESP system on the driving simulator basically produce interventions comparable to those on the real road, and that the effects of varied vehicle parameters and loading conditions upon the interventions of the systems could also be shown in a realistic manner. Yet, for a complete evaluation of the ESP system on the simulator, the simple code used for the first test is not sufficient. Especially for interventions of the ESP system during fast lane change maneuvers, a more precise representation of the dynamic pressure build-up of the pneumatic braking system is necessary. Furthermore, it should be possible to compare the functionality of the ESP system before and after the parameter learning process and also the interaction of the ESP system with the vehicle drive train. For this, it will be necessary to integrate the complete system functionality, which will be done by implementing a C code provided by the system supplier which is compiled on the same operating system platform as used on the driving simulator.

4 Summary, Future Work

For the development of electronic stability systems for trucks, it is useful to integrate vehicle dynamics simulations into the development process to be able to prove the correct system functionality for a wide variety of vehicle variants and driving situations. For this purpose, a software-in-the-loop environment has been established which integrates ESP software codes generated by the brake system supplier into a vehicle simulation interface. Additionally, it was investigated if a transfer of these simulation models and the ESP codes to the driving simulator could give additional benefit for the ESP system development.

During a first test of a truck ESP system on the Daimler driving simulator, it could be shown that the combination of vehicle models with a software-in-the-loop code of the ESP system is able to realistically reproduce the stability interventions of the system and thus can principally be used for optimizing the system for a wide variety of vehicles and vehicle conditions. In the next step, a complete supplier code of the truck ESP will be integrated into the vehicle model as a user routine in order to achieve a complete representation of the system functionality on the driving simulator.

References

1. Baake, U.: Industrielle Nutzfahrzeugentwicklung – Die richtige Spur finden und halten, vol. 4. GFFT Jahrestreffen, Frankfurt (2010)
2. DaimlerChrysler, A.G.: Relevanz von ESP zur Reduzierung von Unfallzahlen. Pressemitteilung (2002)
3. ECE Regulation: Braking (passenger cars). Regulation No. 13-H Version 11 (2009)
4. ISO Standard: Road vehicles – Heavy commercial vehicles and buses – Steady-state circular tests. ISO 14792 (2003)
5. ISO Standard: Road vehicles – Heavy commercial vehicles and buses – Lateral transient response test methods. ISO 14793 (2003)
6. Wüst, K.: Simpack Real-time Models for HiL Testing at Daimler Trucks. In: Proceedings of the Simpack User Meeting (2007)
7. TNO Automotive: MFTYRE & MF-SWIFT 6.1.1 – Release Notes (2008), www.tno.nl
8. Baake, U.: Industrielle Nutzfahrzeugentwicklung in globalen Märkten. Vorlesungsskript RWTH Aachen (2010)

Secure Beamforming for Weather Hazard Warning Application in Car-to-X Communication

Hagen Stübing[1] and Attila Jaeger[2]

[1] Adam Opel GmbH
Active Safety Systems
Friedrich-Lutzmann-Ring
65423 Rüsselsheim, Germany
hagen.stuebing@de.opel.com

[2] Technische Universität Darmstadt
Integrated Circuits and Systems Lab
Hochschulstraße 10
64289 Darmstadt, Germany
jaeger@iss.tu-darmstadt.de

Abstract. Intelligent networking of cars and infrastructure (Car-to-X, C2X) by means of dedicated short range communication represents one of the most promising attempts towards enhancement of active safety and traffic efficiency in the near future. Nevertheless, as an open and decentralized system, Car-to-X is exposed to various attacks against security and driver's privacy. This work presents an approach for enhancing security and privacy on physical layer, i.e. already during sending and receiving of messages. The technique is called *Secure Beamforming* and is based on the radiation patterns produced by the antenna array proposed in [1].

In this work we evaluate the feasibility of this antenna for *Weather Hazard Warning*, a C2X application which includes communication scenarios among cars and between cars and infrastructure. By means of a dedicated simulator, appropriate beams are explored and beamforming protocols for different communication scenarios are proposed.

Keywords: Car-to-X Communication, Secure Beamforming, Weather Hazard Warning.

1 Introduction

Modern vehicles include a multitude of highly sophisticated technologies for active safety. Driver assistance systems like *Adaptive Cruise Control* (ACC), *Active Braking*, *Night Vision* or *Pedestrian Recognition* are only some examples which are expected to enhance future road safety significantly. Basically, all these systems rely their decisions on information received from their local sensors. Among others, these are Sonar, Radar, Lidar, or Camera. For detection of vehicles in

A. Biedermann and H. Gregor Molter (Eds.): Secure Embedded Systems, LNEE 78, pp. 187–206.
springerlink.com

direct proximity, *Park Assistant Systems* use ultrasound sensors which possess a very limited transmission range about 4 m. For detection of more distant vehicles or pedestrians in case of Active Braking, camera-based image detection is deployed. With automotive cameras available today, a reliable detection can be achieved up a total distance of 80 m. Long-range Radar possesses very promising transmission ranges up to 200 m which enables active braking even at high speeds. Compared to Radar, a Lidar-based system can also be deployed to detect heavy rain or snowfall, which delivers valuable input for dynamically calibrating sensitivity of the E*lectronic Stability Control* (ESC) or *Traction Control System* (TCS).

Nevertheless, these sensors generally monitor only the near area around the vehicle, and do not distribute the collected information to other vehicles. Thus, sensing is restricted to line of sight, leaving out hidden and unrecognized but possibly relevant vehicles or other obstacles. Furthermore, since Radar and Lidar are still expensive, car manufacturers will only offer this sort of safety technology to buyers of luxury cars in the near future.

Car-to-X Communication offers new possibilities for enhancing active safety and traffic efficiency at a large scale. The term Car-to-X thereby refers to both, cooperative information interchange between cars (C2C), and between cars and infrastructure (C2I) which includes Road Side Units (RSU). Figure 1 illustrates the extended driver horizon by means of Car-to-X, in comparison to the limited range of local sensors. By means of this, warnings of potential risky situations are no longer limited to local detection only. Instead, a dangerous situation is detected only once and then forwarded via several hops, such that approaching vehicle drivers may react in time and adapt their driving behavior accordingly.

Fig. 1. Vehicle sensor range comparison

The European Federal Commission has dedicated the 5.9 GHz frequency exclusively for transportation-related communications that can transmit data over distances of up to 1000 m. For road safety applications a bandwidth of two 20 MHz contiguous channels is required. One channel is dedicated for network control and safety purposes and the second for safety applications only [2]. C2X communication is based on the IEEE 802.11p standard and enables time critical safety applications at very low data transmission delay. Currently developed chipsets allow communications between stations in ad-hoc mode without the necessity of a permanent infrastructure connection.

Car-to-X enables various kinds of promising applications from different categories. With respect to road safety, *Electronic Brake Light*, *Collision Avoidance on Intersections* or *Weather Hazard Warning* are use cases which may avoid one of the most frequent causes of accidents, today. Traffic management systems may react on traffic events more dynamically because mobility data of all vehicles is constantly monitored by RSUs. Traffic assessment becomes more precise and thus, alternative route management becomes more efficient.

Especially for use cases such as Weather Hazard Warnings, the driver may greatly benefit from the exchange of this foresighted traffic information. The message distribution thereby may include the forwarding via several intermediate stations using Car-to-Car and Car-to-Infrastructure communication links.

Besides, enhancing safety on the road, collecting weather relevant data via a vehicle's local sensors and forwarding of this information via C2X communication is also of high interest for meteorological institutions. For them, vehicles on the road may be regarded as a network of sensor nodes, providing valuable information about the current weather or climate in a specified region. Actually, today's vehicles are equipped with a bundle of sophisticated sensors such as temperature sensors, light sensors, or rain sensors. By distributing sensor probes of all vehicles, a central infrastructure facility may aggregate this data to derive weather relevant information. Including weather information from conventional sources like satellite observations or weather measurement stations, helps metrological institutions to create a more accurate weather view, and thus, enhance their forecasts significantly.

Despite all benefits, Car-to-X applications will contribute to active safety or traffic efficiency they are highly vulnerable towards possible attacks. A comprehensive attack analysis is given by [3]. Especially applications such as warning of aquaplaning or black ice require instant driver reaction and therefore have to be reliable by all means. Corrupted or forged message not only lead to low acceptance of Car-to-X technology by the customer but may also have fatal consequences for road safety. Table 1 summarizes the main threats relevant for Car-to-X and how they are currently addressed by running field trials and draft standards.

Data integrity as well as sender authentication are ensured by means of a *Public Key Infrastructure* (PKI). Digital signatures are used to detect message manipulations, whereas the public keys used to create the signatures are certified by a central *Certification Authority* (CA). Eavesdropping of messages and

Table 1. Possible threats and coutermeasures

Threats	**Countermeasures**
Message Manipulation	Cryptographic, asymmetric signatures based on Elliptic Curves Cryptography (ECC)
Message Forging	Certification of public keys and certain attributes by trusted Public Key Infrastructure (PKI)
Message Replay	Timestamps and/or sequence numbers plus geostamps
Message Falsification	Mobility data verification and plausibility checking
Denial of Service	Monitoring protocols, **Secure Beamforming**
Privacy Infringement	Changing pseudonym identifiers, **Secure Beamforming**

replaying them at different locations may be impeded by including a secure timestamp and geostamp into the signature. In IEEE 1609.2 [4] standard secure message formats and cryptographic algorithms are defined. This standard is currently deployed and further investigated by large scale operational field tests like sim^{TD} [5] and Pre-DriveC2X [6].

An attacker, which is in charge of a valid certificate, is hardly detectable, unless he is sending non-plausible data. In [7] a component has been developed which is capable of verifying the mobility data, i.e. position, speed, and heading inside a message to detect, if the vehicle is following a realistic driving behavior. If any deviating values are observed, the message is marked as insecure. These checks may be further complemented by additional plausibility checks on applications layer. For instance, a message indicating a black ice warning, while the vehicles local temperature sensor measures a temperature above 30 °C, gives evidence on an untrustworthy message.

In this work, we present the deployment of a countermeasure, which may be applied to cope with road side attackers, that cannot be detected by means of cryptographic primitives, e.g., *Denial of Service* attacks. This technique is called *Secure Beamforming* and has been introduced in [8], [9] and [1]. Secure Beamforming refers to applying directional antennas by which one may force transmission energy into desired directions, leaving out the road sides, where potential attackers may be located. In [1], an appropriate simulation-based approach was presented that supports an antenna designer in specifying an antenna aperture, which satisfies security needs. Besides enhancing security level of C2X, this approach may also be applied for enhancing driver's privacy.

While in [9] a comprehensive study on a large set of C2X use cases has been carried out, in this work we will focus on a single use case and perform an in-depth analysis, which communication scenarios may occur within this use case and how Secure Beamforming is designed for each. As one of the most promising use cases, we are investigating on *Weather Hazard Warning*. This use case will have an effect on both, traffic efficiency and road safety, and therefore possesses high security requirements. Furthermore, the detection, forwarding, and aggregation of weather data requires various communication scenarios, including C2C as well as C2I communication, which makes this use case a good candidate to explore the possibilities of Secure Beamforming.

This work is structured as follows: After this introduction, Sect. 2 presents the principles of Secure Beamforming, including the underlying antenna model and attacker assumptions. In Sect. 3, an intensive survey over the Weather Hazard Warning application is given. The Car-to-X architecture is presented, as it is composed of the individual components and communication links between them. In Sect. 4, a Secure Beamforming configuration is derived for each of these communication links and appropriate protocols for scheduling the beams according to the actual traffic scenarios are proposed. Section 5 summarizes the results and gives an outlook on future work.

2 Secure Beamforming Concept

Currently running field operational tests like sim^{TD}[1] or Pre-DriveC2X[2] entirely rely either on single whip antennas or shark fin antennas. These antennas possess a rather omnidirectional radiation pattern, distributing information into all directions uniformly. While being cost-effective and easily realizable, such antennas are highly dissipative concerning channel usage. Such a characteristic is disadvantageous not only for channel capacity reasons. As illustrated in Fig. 2(a) an omnidirectional pattern furthermore allows an attacker to operate very far away from the road side.

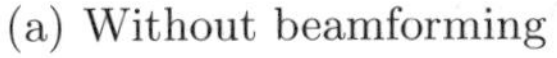
(a) Without beamforming

(b) With beamforming

Fig. 2. Attacker exclusion by Secure Beamforming

In [9] we actually examined that for almost all use cases, the message validity is never omnidirectional. In fact, a message generally possesses a geographical

[1] www.simtd.de

[2] www.pre-drive-c2x.eu

validity, i.e. the message is either relevant for vehicles located to the front, to the back or to the side, in case of a RSU connection. By mapping the radiation pattern for the message to its geographical validity, an attacker on the road side, as shown in Fig. 2(b), will have to put much more effort into his antenna system to still receive the message with a sufficient power level. In conjunction with frequently changing pseudonyms this technique impedes tracking of vehicles considerably.

Furthermore, adapting the antenna characteristic not only applies for the sending of messages. Instead, a radiation pattern also defines which directions are amplified and which are suppressed for the receiving case. By orienting the receiving pattern towards relevant senders, leaving out the road sides, security on a physical layer is enhanced. Please note that Secure Beamforming is a complementary technique and as such it is embedded inseparably into an overall security solution as depicted in previous sections.

2.1 Attacker Model

In this work a very sophisticated attacker is assumed, who is in charge of sending plausible and correctly signed messages with valid certificates. Thus, a forged message cannot be detected by cryptographic primitives. In this case, Secure Beamforming represents a way to suppress the messages sent by this roadside attacker.

Furthermore, the attacker is equipped with a common C2X communication system, consisting of a whip antenna and the corresponding 802.11p transceiver. Located about 400 m away from the road, e.g., on a parking place or nearby housing area, he is trying to inject forged weather messages and to eavesdrop exchanged messages.

2.2 Antenna Model

The total field of linear antenna arrays results from the vector addition of the fields radiated by the individual elements. To provide very directive patterns, it is necessary that the fields from the elements of the array interfere constructively in the desired directions and interfere destructively in the remaining space. Assuming equidistant antenna elements and an identical feeding current, the overall pattern is shaped by at least three parameter [10]:

1. the number of antennas N,
2. the distance d between the individual elements,
3. the excitation phase α of the individual elements.

According to a specified configuration, a resulting radiation pattern is produced, which is described by the pattern of an individual array element multiplied by the Array Factor AF:

$$AF = \frac{1}{N} \cdot \left(\frac{\sin(\frac{N}{2}\Psi)}{\sin(\frac{\Psi}{2})} \right) \tag{1}$$

Within this Array Factor N denotes the number of elements in that array and Ψ is the progressive angle of the antenna array. For a linear array, the progressive angle is a function of the element separation d, the phase shift α, the wave number $k = \frac{2\pi}{\lambda}$ and elevation angle θ:

$$\Psi = \alpha + kd \cdot \cos(\theta) \tag{2}$$

For all simulation processes a wavelength $\lambda = 5.1688\,\text{cm}$ was selected, which corresponds to the C2X communication frequency of 5.9 GHz. By choosing the distance d between the individual elements as multiple of the wavelength, the array factor becomes independent from the frequency.

Our antenna model consists of two orthogonal antenna arrays which are steered independently from each other as proposed in [9]. By means of this antenna system a great variety of different configurations in terms of type, number, distance, and phase shift of antennas is given. In order to yield a specified pattern characteristic, according to the array factor, only some elements of the antenna array in Fig. 3 are biased. The transmission power for each packet may be varied individually with a grading of 0.5 dB from 0 dBm to a maximum of 21 dBm.

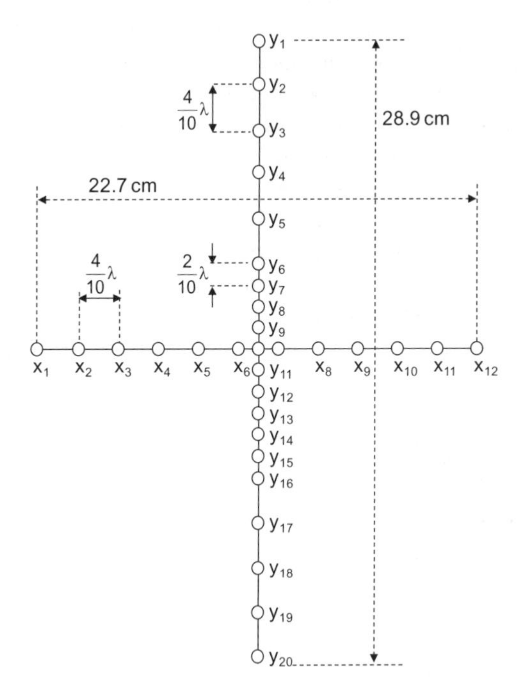

Fig. 3. Antenna model

3 Weather Hazard Warning Application

One of the most promising applications for C2X communication is *Weather Hazard Warning.* This application informs the driver about upcoming local weather events which may influence road safety so that he may adapt his driving behavior on time and pass hazardous driving situations unharmed.

In this section, we describe the setup for such an application. Since realization of the Weather Hazard Warning application highly depends on available communication links and components inside the C2X architecture, we describe a most likely C2X system architecture that enables such a Weather Hazard Warning application. These application assumptions are used to develop feasible beamforming strategies in the following section.

3.1 Information Distribution Requirements

Currently, a large infrastructure of approved and well equipped weather measurement stations is established. However, knowledge about road safety relevant weather information is present in weather services like *Deutscher Wetter Dienst* (DWD, a German meteorological institute) but are rarely available to drivers on the road.

On the other side, a large number of vehicles are widely spread all over the roads. Vehicles already carry sensors to measure weather relevant data, e.g. temperature or rain fall. This data heavily extends knowledge delivered by established weather stations, if they were available to weather services.

Distributing weather warnings to relevant vehicles was proposed in [11] and usually realized via *Traffic Message Channel* (TMC). But this technology is not yet available in large scale and moreover, information transmission is only unidirectional. However, a Weather Hazard Warning application in C2X communication requires information distribution in both directions. Accordingly, a bidirectional communication link between vehicles on one side and weather measurement stations as well as weather services on the other side has to be established.

3.2 Weather C2X Architecture

Actual, there is no standardized C2X architecture available. We therefore rely our system assumptions on architectures developed in large scale field operational tests [12], consortiums [13], and standardization committees [2]. The architecture and how our Weather Hazard Warning is embedded is described briefly in this section.

Basically every entity involved in C2X has to be equipped with an *On Board Unit* (OBU), which consists of equipment for wireless communication (*Control Communication Unit*, CCU) and a processing unit (*Application Unit*, AU), e.g., a small personal computer on which functionality for use cases is realized. This distinction is rather logical and does not imply physical distinction. Therefore,

CCU and AU may be in just one physical device. Moreover, the separation between these components is not well-defined and may vary in different implementations. Vehicles equipped with C2X technology furthermore require a *Human Machine Interface* (HMI).

Beside vehicles, also *Road Side Units* (RSU) are involved and distributed along the roads. These stations are used to store and forward messages if no direct vehicle to vehicle communication is possible. For that reason, RSUs communicate via several communication technologies. On one hand, messages are exchanged to vehicles via IEEE 802.11p wireless communication. On the other hand, RSUs serve as a communication interface to wired backend network, enabling high data transfer over large distances. This network interconnects RSUs to each other and to local *Traffic Management Centers*, e.g., Hessian Traffic Center (a local road authority in the state of Hessian, Germany). Since Traffic Management Centers typically are already connected to infrastructure facilities, service providers, and the internet, vehicles have also access to these via RSUs.

To realize the Weather Hazard Warning application, we introduce the *Road Weather Center* (RWC), as new part of the Traffic Management Centers. The RWC connects all parties involved in weather warnings and gathers data from local weather services, measurement stations, and vehicles. This gathered weather data is used to generate highly reliable weather warnings based on a more comprehensive view on the actual weather situation. RWC furthermore forwards weather warnings and recorded sensor measurements from vehicles to weather services to improve their knowledge as well.

3.3 In-Vehicle Weather Detection

Actual vehicles contain several sensors to gather weather relevant data. However, these sensor data may be complemented by observing driver's action, e.g., switching on front shield wipers or rear fog light.

All this information is used to detect hazardous weather events. In our use case we distinguish three different weather events which are detected. In Table 2 these events and associated sensors and driver actions are listed. A more detailed overview on how sensors may be used is given in [14].

In general, a single vehicle data is not sufficient for reliable weather event detection. Vehicle data must be weighed against each other and evaluated in context of the current driving situation. For instance, ESC not only reacts on slippery roads but also, and probably more often, on rapid driving through curves. Table 3 very briefly lists proposals how to combine vehicle data to get event indication. Thereby a focus is set to widely available vehicle data. Rarely available sensors are intended to validate or support these data.

For all types of events an indication has to be active for a minimum amount of time to overcome short appearance of runaway values or accidentally pushed buttons by the driver. In case of leaving an event same criteria are applied.

Table 2. Weather events with associated vehicle data

Weather Event	Hazard	Indicating Vehicle Data
Fog	reduced visibility	– fog lights activity – vehicle speed – air humidity – Radar or Lidar used in ACC
Heavy Rain	reduced visibility and aquaplaning	– front shield wiper speed – vehicle speed – rain sensor
Black Ice or Frozen Snow	slippery road	– exterior temperature – vehicle speed – anti-lock braking system (ABS) – ESC

Table 3. Proposals how to indicate weather events

Weather Event	Vehicle Data Combination
Fog	Rear fog light is active and vehicle speed is below 50 km/h. Air humidity and radar may be used to validate fog light usage.
Heavy Rain	Wiper speed is set to level two and vehicle speed is below 80 km/h Rain sensors may deliver more authentic level of rainfall than wiper speed.
Black Ice or Frozen Snow	Exterior temperature is below 2 °C, vehicle speed is below 50 km/h, ABS is active even on low brake intensity. Unexpected ESC activity in large curves or strait roads can be considered in addition.

3.4 Event Notification Strategies

To communicate a weather event to other road participants, vehicles are sending short notification messages when first detecting an event. While driving through the hazard vehicles periodically update this notification. When leaving an event, a final notification is sent to mark the hazardous zone. These notifications are sent, even if the vehicle already received a notification from other road traffic participants or the RWC. This way it confirms und updates the older notifications and a more precise and reliable database can be generated.

The driver gets warnings on hazardous weather events via the HMI. This notifications and warnings may be visual, acoustical or haptical. Drivers should get warnings on weather events before they reach the hazardous zone so that they may adapt their driving behavior on time. Therefore each notification message contains three areas:

1. *Hazard Zone*: The zone where the hazardous weather event is located.
2. *Notification Area*: The area, in which the driver should be informed about the hazardous event. This area typically is larger as the Hazard Zone.
3. *Dissemination Area*: The area, in which the messages must be distribued. This area typically is larger as the Notification Area.

Displaying a warning after the driver has already reached the hazardous zone is not reasonable, so this has to be avoided. Furthermore, a driver acting in expected behavior should not be warned, because he does not need to adapt his behavior. If a driver does not react, warnings become more insistent till reaching the hazardous zone.

3.5 Involved Message Types

Due to different communication scenarios, there are two different message types involved in this application. These are respectively, the *Decentralized Environmental Notification* (DEN) and the *Probe Vehicle Data* (PVD). The previous mentioned CAMs are not involved in the Weather Hazard Warning.

DENs are used as explicit notification of a dangerous weather event in the near environment. It may be generated either from a RWC or directly from vehicles driving nearby. A DEN mainly consists of event type, location, detection time and message validity time. Thereby, the location usually is encoded as a rectangle surrounding the hazardous area. The type gives evidence on the weather hazard.

PVD consist of collected sets of probe values with timestamp and geographical position. These sets are gathered while the vehicle is driving along the road and accumulated in order to build up a *weather profile* for the past region. The weather profile contains sensor measurements like temperature, air humidity and barometric pressure. This enables conclusions on the weather situation.

In contrast to DENs, PVD are rarely processed data and need to be interpreted. Due to rather low processing power in vehicles, this extensive data evaluation is not performed on the vehicles, but on infrastructure side. Hence PVD are not transmitted to other vehicles but are forwarded to RWC.

3.6 Data Aggregation and Fusion

Since vehicles receive DENs from multiple sources, comparing and aggregation of this data has to be performed. Furthermore received messages have to be compared with own vehicle sensor data.

However, since a RWC gathers also PVD, measurements of weather stations, and information from weather services, data fusion has to be done with all this different kind of information. Weather stations deliver highly precise measurements from various sensors not available in vehicles, e.g. dew-point temperature, air temperature, street and in earth temperature, sight distance, rainfall intensity, and remaining salt ratio on the street. So vehicle generated warnings can be confirmed and improved for areas with these measurement stations. Weather station measurements are valid for local area only. RWC combine these measurements with weather profiles to extract an overall weather view for the entire region.

Furthermore weather stations have sensors for wind speed and direction. So the RWC is able to generate warnings for strong crosswinds, which are not detectable by means of local vehicle sensors.

3.7 Communication Scenarios

As described above, in Weather Hazard Warning application, different message types are transmitted between vehicles and between vehicles and infrastructure. This results in different communication scenarios as described in the following.

Weather Hazard Notification scenario includes sending of DENs, which are sent from vehicles to other vehicles and also to RWCs via RSUs, as indicated in Fig. 4.

The second scenario refers to *Weather Hazard Forwarding* and uses *Store & Forward* mechanism according to [13] to distribute warnings over larger distances. In Fig. 4 forwarding over three hops and different lanes is depicted. While forwarding, vehicles only send messages into opposite direction of recieving, i.e., away from the Hazard Zone.

The last scenario is called *Probe Data Distribution.* This is transmission of PVDs from vehicles to the RWC via RSUs. It has to mentioned, that PVDs are large messages, therefore the relative position between vehicle and RSU will vary considerable during transmission. The exchange of large PVD messages between vehicles is as afore mentioned not intended.

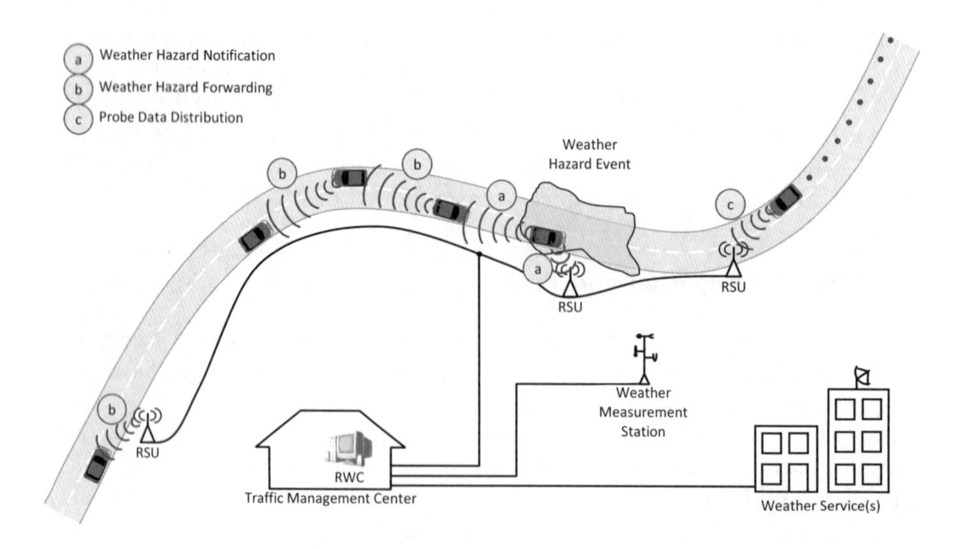

Fig. 4. Overview of Weather Hazard Warning communication scenarios

4 Simulation

In this section, we will derive an appropriate pattern steering of our antenna model for each of the communication types occurring inside the weather warning application. To apply the approach of beamforming we developed a simulator [1], which works on top of MATLAB, a well-known simulation environment in the automotive domain. Many built-in routines for mathematical calculations in MATLAB were used to yield an efficient calculation of the antenna field.

4.1 Simulation Objectives

From the previous description of Weather Hazard Warning applications, we may recall three basic communication scenarios. These are respectively: *Weather Hazard Notification*, *Weather Hazard Forwarding*, and *Probe Data Distribution*. Before starting the simulation process the required radiation patterns have to be defined for each of the three communication scenarios. In the following we will do this by means of argumentative reasoning.

Before starting the simulation process the required radiation patterns have to be defined for each of the three communication scenarios. In the following we will do this by means of argumentative reasoning.

If a vehicle detects a whether hazard such as black ice, heavy rain or fog by its local sensors it instantly distributes a corresponding warning message to all surrounding vehicles. By nature, weather hazards are not limited to the vehicles positions only, but rather have an effect on the entire area around the vehicle. Consequently, these messages are relevant for all vehicles in communication range. An a priori dissemination direction is not anticipated for the *Weather Hazard Notification* communication scenario. Nevertheless, a beamforming adaption according to the position of vehicles in communication range is intended.

Via Store & Forward techniques according to [13] hazard warning messages are forwarded over longer distances to warn approaching traffic in time. For *Weather Hazard Forwarding* the geographical validity is similar to the previous communication scenario. Besides adapting the field to surrounding vehicles, further refinements may be applied according to the intended dissemination direction of the message. Assuming that messages sent by the originator have also been received by all surrounding vehicles, the next forwarder may adapt his radiation pattern in a way that already notified vehicles left out.

The *Probe Data Distribution* refers to the use case of a *Local RSU Connection*, and therefore possesses similar requirements concerning the pattern steering as described in [9]. The exchanged data does not only include probe of all sensors, relevant for weather identification but further includes related information for other applications, such as congestion warning. This requires the exchange of large sets of data when passing by a RSU. In order to support the secure transmission of these packets, a mutual beamforming of both, sender and receiver is anticipated.

In the following section, an appropriate beamforming for each communication scenario is determined. The related antenna configurations are stated, and beamforming protocols for scheduling of the different beams are defined.

4.2 Secure Beamforming Design Methodology

The deployed design methodology refers to the simulation-based approach proposed in [9]. Accordingly, the designer determines the respective antenna configuration via graphical comparison of the required pattern with achieved radiation pattern produced by the beamforming simulator. In the following the steps as defined in [1] are performed:

1. **Defining the attacker model**: In the scope of this work we limit our analysis to static attackers located on the roadside as described in Sect. 2.1.
2. **Selecting the scenario**: The scenarios are chosen accordingly to Sect. 3.7, reflecting the most common communication scenarios for Weather Hazard Warning application:
   (a) Weather Hazard Notification,
   (b) Weather Hazard Forwarding and
   (c) Probe Data Distibution.
3. **Selecting the traffic and road situation**: In this work, beamforming for weather applications is investigated with respect to geometry of motorway and rural roads.
4. **Configuring the antenna array**: Compared to [9], in this work the design space exploration is limited to configurations achievable with the antenna model stated in Sect. 2.2.
5. **Evaluating the generated patterns**: Section 4.1 gives evidence on the patterns, required for each communication scenario. To find a matching pattern, the system designer explores all achievable patterns by varying the different configurations for the antenna model stated in Fig. 3. In case of conformance, the corresponding configuration is recorded along with the respective communication scenario, as defined during step 2.

Please note that this design flow supports the designer in defining which antennas are biased within the antenna model. Hence, the phase shift inside the feeding signal is adjusted according to the relative position of the receiver station and may be determined analytically.

4.3 Simulation Results

Figure 5 summarizes the set of different beams, feasible for the Weather Hazard Warning application. In Fig. 6 the corresponding configurations of our antenna model are stated. For each configuration only a specified subset of antenna elements is biased.

Apparently, the beams produced by the antenna model fulfill the requirements stated in the previous Sect.4.1. According to the position of vehicles in reception range the main beam is steered into the respective directions.

Pattern Fig. 5(a) restricts message dissemination and reception to regions located along the same lane as the host vehicle. By translating the driving angle into a related phase shift of the feeding signal, the pattern is oriented into driving direction. That way, high directivity is ensured even when driving curves as depicted in Fig. 5(b).

Biasing antenna elements y1-y16 and applying a phase shift of 207° produces a very narrow cone towards the back of the vehicle (Fig. 5(c)). In comparison, a broadside pattern is chosen to sent messages to all vehicles driving to the front of our host vehicle (Fig. 5(d)). The Road Restricted Pattern in (Fig. 5(e)) refers to the omnidirectional case, where all vehicles on the road are addressed. For unicast messages to RSUs a very directive pattern is produced and steered into the respective direction (Fig. 5(f)). Please note that the presented patterns may be used for sending or receiving equivalently.

Generally, it is recommended for vehicle to select an appropriate beamforming in a way that minimum area is covered and in contrary no safety relevant message gets lost. In the following sections a simple beamforming protocol is proposed, dealing with that issue.

4.4 Beamforming Protocols for Wheather Hazard Warning

Selecting and steering the different beams is a task performed on lower abstraction layers of the ITS reference model [2]. In Fig. 7 and Fig. 8 the cross-layer protocol between application and MAC layer is stated.

If a weather hazard message is handed over for sending, a beamforming component on MAC layer analyses the message header to determine the communication scenario according to the previous section. Car-to-X architectures like [12] generally foresee neighborhood tables, which include the frequently updated mobility data of all vehicles in communication range. We specify a *relevance filter* to extract those vehicles from the neighborhood table, which are located inside the intended message dissemination area. For the *Weather Hazard Notification* scenario the dissemination area is considered omnidirectional, which includes all stations inside the neighborhood table. In case of *Weather Hazard Forwarding* scenario, the relevance filter is set such that all stations are extracted which are located inside the message dissemination direction. Please note that the dissemination direction is derived from the dissemination area, specified inside the sent message as described in Sect. 3.

As the next step, an appropriate beam pattern is selected from the set of feasible beams stated in Fig. 5. By adjusting the power level of the feeding signal, the communication range is scaled to the most distant vehicle.

Probe Vehicle Data is sent every time the vehicle passes by a Road Side Unit. For that purpose, the position-based pattern (Fig. 5(f)) is deployed to produce a very narrow beam towards the position of the RSU. To steer that beam according to the dynamic driving behavior of the vehicle, the relative geometric angle between both stations is calculated and transferred to an electrical phase shift of the feeding signal. Furthermore, mutual beamforming between both stations requires a handshake protocol at start-up.

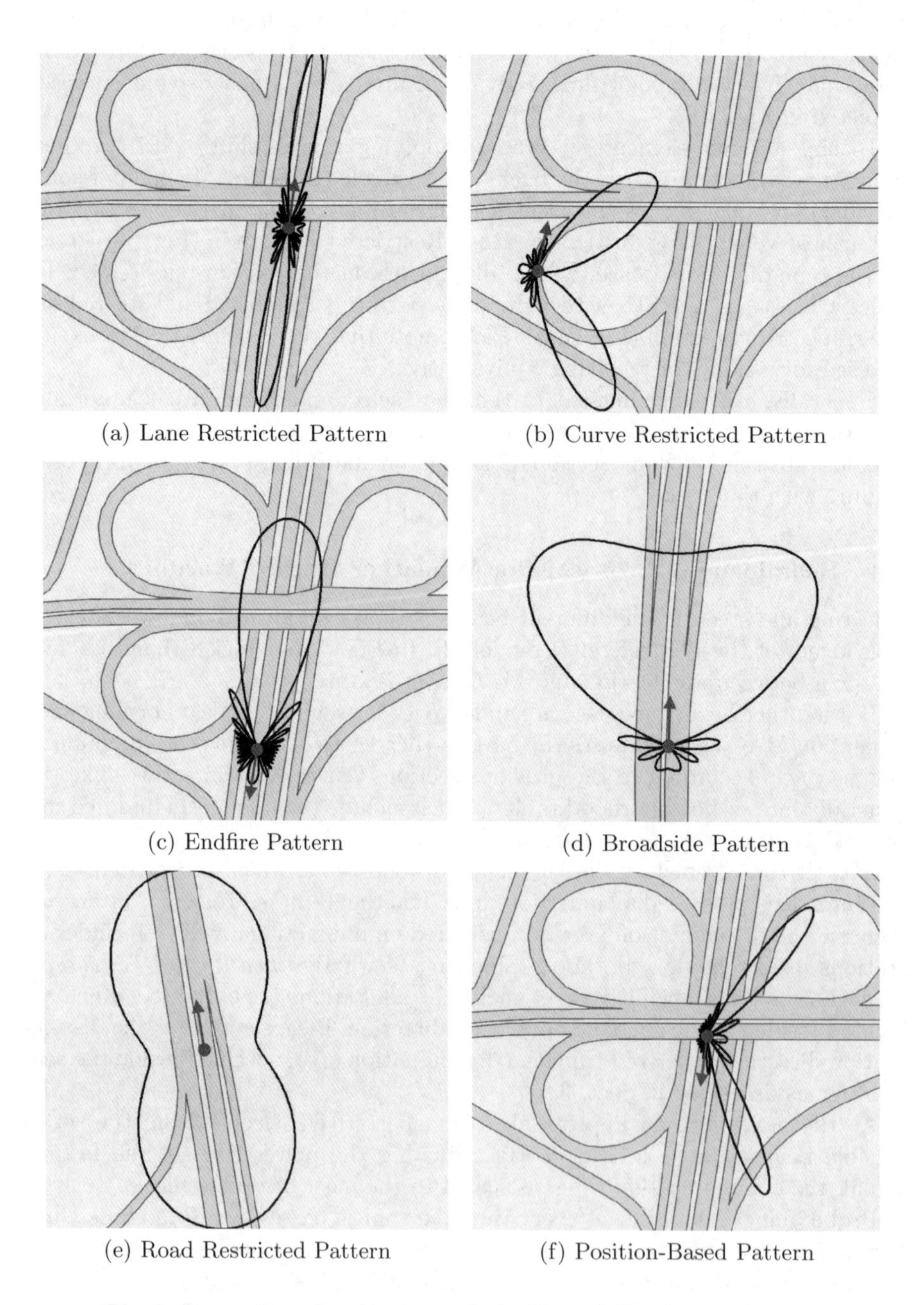
(a) Lane Restricted Pattern
(b) Curve Restricted Pattern
(c) Endfire Pattern
(d) Broadside Pattern
(e) Road Restricted Pattern
(f) Position-Based Pattern

Fig. 5. Secure Beamforming patterns for Hazard Weather application

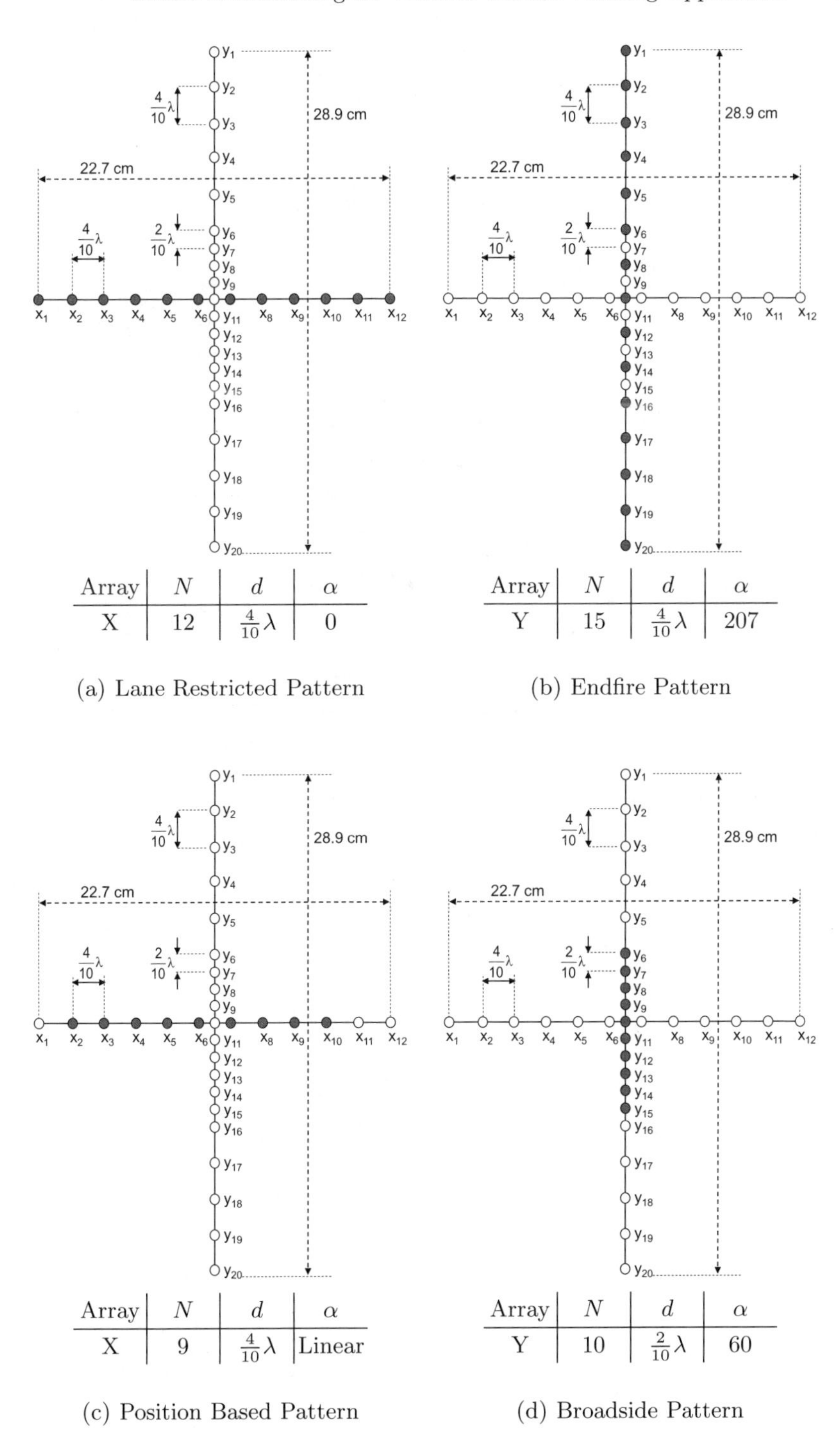

Array	N	d	α
X	12	$\frac{4}{10}\lambda$	0

(a) Lane Restricted Pattern

Array	N	d	α
Y	15	$\frac{4}{10}\lambda$	207

(b) Endfire Pattern

Array	N	d	α
X	9	$\frac{4}{10}\lambda$	Linear

(c) Position Based Pattern

Array	N	d	α
Y	10	$\frac{2}{10}\lambda$	60

(d) Broadside Pattern

Fig. 6. Antenna array configurations

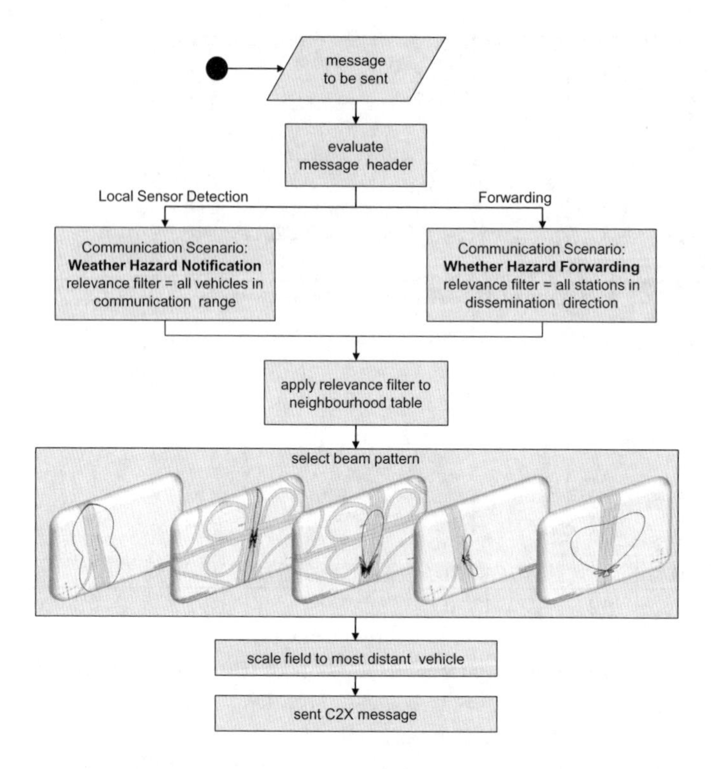

Fig. 7. Beamforming protocol for Weather Hazard Warning and Weather Hazard Forwarding

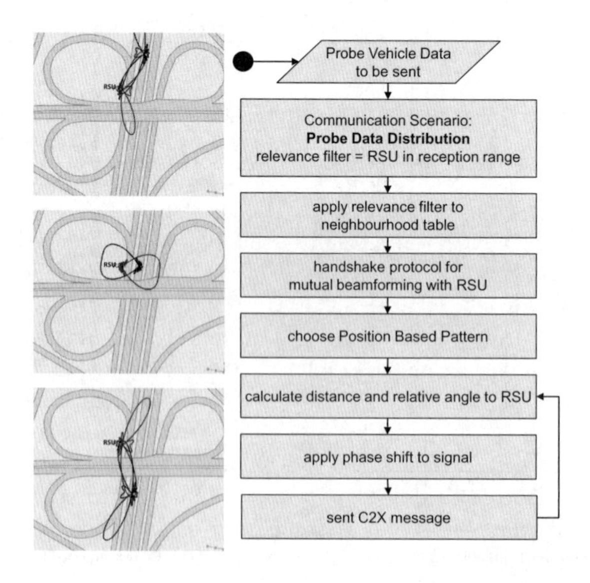

Fig. 8. Mutual Beamforming protocol for Probe Vehicle Dissemination

The proposed secure beamforming protocols serve the purpose of enhancing the driver's privacy by minimizing the dissemination area of messages. To enhance security, similar protocols have to be found for the receiving case. Whereas the beamforming for sending is bounded to a single message, the pattern for receiving has a large impact on all kinds of messages received. A dynamic arbitration of the channel, depending on the requirements of the different applications is subject of future work. In this work we propose to use road-restricted pattern according to Fig. 5(e) for Weather Hazard Warning.

5 Summary and Conclusion

In this work, the feasibility of secure beamforming for Car-to-X has been studied by means of a representative use case. We have decided to perform our analysis on Weather Hazard Warning, because it is considered as one of the most sensitive use cases towards possible attacks. Instant driver reaction is required in case of black ice warning and therefore underlies high security needs.

A detailed survey of the Weather Hazard Warning has been given, starting from the event detection, message generation to the final distribution and forwarding via several hops. The description of this use cases refers to the full specification of *Road Weather Warning* and *Identification of Road Weather* according to [15].

We identified three different communication scenarios within this Weather Hazard Warning use case: *Weather Hazard Notification*, *Weather Hazard Forwarding* and *Probe Data Distribution*. For those scenarios the requirements regarding beamforming are postulated. From that we concluded that geographical validity of messages is hardly ever omnidirectional, but may be restricted to the regions where intended receivers are located.

In order to find a related beamforming which matches the geographical validity for the different scenarios, a simulation-based approach has been applied. Several traffic situations with varying vehicles densities and road types have been modeled by means of a dedicated simulator. By switching different configurations of the anticipated antenna model, the appropriateness of the resulting pattern has been evaluated. In fact, a large set of the available patterns with this antenna model are feasible for the examined Weather Hazard Warning application. Especially for a rear-oriented as well as forward-oriented dissipation of messages the benefits of beamforming become evident. The antenna may produce a radiation pattern, which covers only the same road in both directions, limiting the area from where an attacker may operate. In order to exchange larger sets of PVD, a mutual beamforming of vehicle and RSU has been proposed.

The scheduling of different beams is a non-trivial task, as it is highly dependent on a precise traffic assessment and correct decisions, which vehicles are relevant as receivers. We presented two beamforming protocols by which the relevant recipients are determined with respect to the intended dissipation direction.

In our future work, we will adapt and further refine our beamforming concepts by means of real-world measurements obtained from field operational test sim^{TD}. Especially the interoperability of beamforming protocols from different and concurrently running applications represents an important perquisite for successful deployment of this novel security technique.

References

1. Stuebing, H., Shoufan, A., Huss, S.A.: Enhancing Security and Privacy in C2X Communication by Radiation Pattern Control. In: 3rd IEEE International Symposium on Wireless Vehicular Communications (WIVEC 2010), Taipei (2010)
2. Standard: ETSI: Intelligent Transport Systems (ITS); Communications; Architecture. ETSI Draft TS 102 665 (2009)
3. Buttyan, L., Hubaux, J.P.: Security and Cooperation in Wireless Networks. Cambridge Univ. Press, Cambridge (2007)
4. Standard: IEEE Vehicular Technology Society: IEEE Trial-Use Standard for Wireless Access in Vehicular Environments – Security Services for Applications and Management Messages. 1609.2^{TM}-2006 (2006)
5. Bißmeyer, N., Stuebing, H., Mattheß, M., Stotz, J.P., Schütte, J., Gerlach, M., Friederici, F.: Sim^{TD} Security Architecture: Deployment of a Security and Privacy Architecture in Field Operational Tests. In: 7th Embedded Security in Cars Conference (ESCAR), Düsseldorf (2009)
6. Gerlach, M., Friederici, F., Held, A., Friesen, V., Festag, A., Hayashi, M., Stübing, H., Weyl, B.: Deliverable D1.3 – Security Architecture. In: PRE-DRIVE C2X (2009)
7. Stuebing, H., Jaeger, A., Bißmeyer, N., Schmidt, C., Huss, S.A.: Verifying Mobility Data under Privacy Considerations in Car-to-X Communication. In: 17th ITS World Congress, Busan 2010 (2010)
8. Stuebing, H., Shoufan, A., Huss, S.A.: Secure C2X Communication based on Adaptive Beamforming. In: 14th VDI International Conference on Electronic for Vehicles, Baden-Baden (2009)
9. Stuebing, H., Shoufan, A., Huss, S.A.: A Demonstrator for Beamforming in C2X Communication. In: 3rd IEEE International Symposium on Wireless Vehicular Communications (WIVEC 2010), Taipei (2010)
10. Balanis, C.A.: Antenna Theory – Analysis and Design, 3rd edn., ch. 6, pp. 283–333. John Wiley & Sons, Chichester (2005)
11. Patent: US 2002/0 067 289 A1: United States of America (2002)
12. Stuebing, H., Bechler, M., Heussner, D., May, T., Radusch, I., Rechner, H., Vogel, P.: Sim^{TD}: A Car-to-X System Architecture For Field Operational Tests. IEEE Communications Magazine (2010)
13. CAR 2 CAR Communication Consortium: C2C-CC Manifesto - Overview of the C2C-CC System (2007), `http://www.car-to-car.org/index.php?id=31`
14. Petty, K.R., Mahoney III, W.P.: Enhancing road weather information through vehicle infrastructure integration. Transportation Research Record: Journal of the Transportation Research Board 2015, 132–140 (2007)
15. Karl, N., et al.: Deliverable D11.1 – Description of the C2X Functions. Sichere Inteligente Mobilität –Testfeld Deutschland (2009)

Author Index